EARLY DIAGENESIS

A Theoretical Approach

PRINCETON SERIES IN GEOCHEMISTRY

Edited by Heinrich D. Holland

EARLY DIAGENESIS

A Theoretical Approach

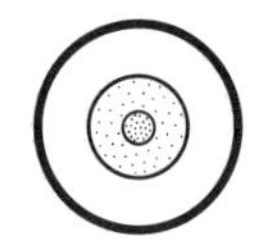

ROBERT A. BERNER

PRINCETON UNIVERSITY PRESS
PRINCETON, N.J.

Published by Princeton University Press, Princeton, New Jersey
In the United Kingdom:
Princeton University Press, Oxford

Library of Congress Cataloging in Publication Data will be found on the last printed page of this book

Clothbound editions of Princeton University Press books are printed on acid-free paper, and binding materials are chosen for strength and durability.

Printed in the United States of America by Princeton University Press, Princeton, New Jersey

Design by Laury A. Egan

9 8 7 6 5 4 3

TO MY WIFE
Elizabeth Kay Berner

Contents

Part II. APPLICATIONS

Preface

When I first started writing this book it was my intention to revise and update my textbook *Principles of Chemical Sedimentology*. However, I soon discovered that so much had transpired in the past ten years in the general area of chemical sedimentology that it would be impossible for me to complete the task during a semester's leave. I first attacked the subject of diagenetic theory and soon found there was more than enough material in this one area alone to make up a separate book on the subject. Besides, it has been one of my major research interests over the past two decades. In this way the present work came into being.

The use of diffusion-advection-reaction models ten years ago was virtually unheard of; today it is commonplace. This has come about partly because of an abundance of interstitial water chemical data which has been collected during the past fifteen years, but also partly because of the increased training of earth scientist graduate students in the basic sciences, which has enabled them to undertake and better understand such modeling. This book is intended for the use of this "new wave" of graduate students and researchers as well as for anyone else who wishes to approach the study of sediments from a quantitative standpoint. (It is amusing to see the variegated collection of labels we apply to individuals who do the kinds of research summarized in this book: sedimentologists, geochemists, oceanographers, marine biologists, limnologists, biogeochemists, environmental scientists, economic geologists, mathematical geologists, even geophysicists.)

I am indebted to the following individuals who generously reviewed the manuscript and offered many helpful suggestions: Robert C. Aller of the University of Chicago, Antonio C. Lasaga of Pennsylvania State University, and Bernard P. Boudreau of Yale University. Yale University provided me with a leave of absence which was sufficient time to complete the manuscript. Finally, I acknowledge the musical inspiration of Maurice Ravel, Camille Saint Saëns, and Joseph Szulc.

New Haven, Connecticut
March, 1980

Robert A. Berner

List of Most Commonly Used Symbols

$\bar{A}$ = surface area of a solid or solids per unit volume of pore solution

a = activity

$\hat{C}$ = concentration in terms of mass per unit volume of total sediment (solids plus water)

$\hat{C}_s$ = concentration of a solid component in terms of mass per unit volume of total sediment

C_s = concentration of a solid component in terms of mass per unit mass of total solids

$\bar{C}$ = concentration of an adsorbed component in terms of mass per unit mass of total solids

C = concentration of a dissolved component in terms of mass per unit volume of pore solution

C_{eq} = saturation equilibrium concentration

C_∞ = concentration out in solution; concentration at depth where $\partial C/\partial x \to 0$ (*asymptotic concentration*)

C_0 = *concentration at* $x = 0$

D = generalized diffusion coefficient; molecular diffusion coefficient in water (area per unit time)

D_s = molecular diffusion coefficient in sediment including the effects of tortuosity (area of sediment per unit time)

D_{wc} = wave and current mixing (diffusion) coefficient

D_B = biodiffusion coefficient for solids

D_I = irrigation coefficient

D_{BI} = biodiffusion coefficient for pore water ($= D_B + D_I$)

F = formation factor (in molecular diffusion); concentration conversion factor ($= \bar{\rho}_s(1 - \phi)/\phi$)

G = concentration of non-diffusable organic matter (expressed as mass of organic carbon per unit mass of total solids) which can be decomposed to CO_2, H_2O, and/or CH_4 via a given bacterial process or set of processes; also Gibbs free energy

J = diffusion flux in general (mass per unit area per unit time)

J_s = diffusive flux in a sediment (mass per unit area of sediment per unit time)

K' = linear adsorption isotherm ($K' = \bar{C}/C$)

K = dimensionless linear adsorption isotherm ($= FK'$); thermodynamic (activity) equilibrium constant

K_c = concentration solubility product constant

k = rate constant (time^{-1})

L = depth (thickness) of the zone of bioturbation

$\mathscr{L}$ = stoichiometric coefficient for sulfate reduction (usually = 1/2)

Q = water flux in a sediment in terms of volume of water per unit area of sediment per unit time

R = rate of a chemical reaction in mass per unit volume per unit time

$\bar{R}$ = rate of chemical reaction occurring on the surfaces of solids

$\mathscr{R}$ = rate of mass sedimentation (mass per unit area per unit time)

r = spherical or cylindrical coordinate

r_c = radius of a spherical crystal, concretion, etc.

t = time

U = generalized water flow velocity (usually lateral) relative to particles

v = velocity of burial of water below the sediment-water interface

v_g = vertical water flow velocity relative to particles

$\mathscr{v}$ = molar volume

x = depth in sediment below the sediment-water interface (sometimes depth below the base of the zone of bioturbation)

y = lateral dimension in a sediment

γ = activity coefficient for a dissolved species

λ = radioactive decay constant

$\bar{\rho}_s$ = mean density of total sediment solids

ρ_w^* = mass of H_2O per unit volume of pore solution

σ = specific interfacial (surface) free energy

ϕ = porosity (connected porosity) in volume of pore water per unit volume of total sediment (solids plus water); also used for total porosity

Ω = saturation state of pore solution (equals ion activity product divided by equilibrium ion activity product (solubility product))

ω = velocity of burial of solid particles below the sediment-water interface

$\left.\begin{matrix}\phi_x\\ v_x\\ \omega_x\end{matrix}\right\}$ = asymptotic values of ϕ, v, and ω at depth where $\partial\phi/\partial x \to 0$

EARLY DIAGENESIS

A Theoretical Approach

Dass am Grunde des Meeres grossartige chemische Processe vor sich gehen, ist kaum in Zweifel zu ziehen. Sie sind es, durch deren Wirkung wir uns die theilweise Umbildung (Diagenese) und Verfestigung des abgesetzten Materials durch Neubildung von Zwischenmassen, wie sie uns bei Meeresablagerungen älterer Zeit entgegentreten, erklären müssen.

K. von Gümbel, *Grundzüge der Geologie*, Fischer, Kassel (1888) p. 334.

1

Introduction

The subject of this book is diagenesis. Diagenesis refers to the sum total of processes that bring about changes in a sediment or sedimentary rock subsequent to deposition in water (sediments deposited in air will not be discussed here). The processes may be physical, chemical, and/or biological in nature and may occur at any time subsequent to the arrival of a particle at the sediment-water interface. However, if the changes occur after contact of the particle with the atmosphere, as a result of uplift, they fall under the heading of weathering, and if they occur after exposure to elevated temperatures upon burial, they are referred to as metamorphism. Often there are problems concerning the boundaries between diagenesis and weathering, and diagenesis and metamorphism, but in this book we avoid them because we will be concerned only with early diagenesis. Early diagenesis, as defined here, refers to changes occurring during burial to a few hundred meters where elevated temperatures are not encountered and where uplift above sea level (or lake level) does not occur, so that the pore spaces of the sediment are continually filled with water. Some examples of early diagenetic processes are: the compactive dewatering of clay-muds, the destruction of lamination by burrowing benthic organisms, the diffusion of dissolved salts in lake beds, the bacterial decomposition of organic matter, the dissolutive removal of calcium carbonate in deep sea sediments, and the formation of concretions.

The subject of early diagenesis can be approached on three levels. The first and simplest is qualitative generalizations based on depth trends observed in actual sediments. An example is the bacterial reduction of sulfate to H_2S which is evidenced in organic-rich, anoxic marine sediments by decreases with depth in the concentration of dissolved sulfate. The second level is qualitative predictions based on laboratory measurements and thermodynamic calculations. For our example, lab studies have shown that in the presence of certain organic compounds, sulfate is reduced to sulfide under anoxic conditions, but only by bacteria. In addition, free energy data enable one to calculate that sulfate reduction accompanying organic matter decomposition should occur during early diagenesis, and that this process should not proceed in the presence of dissolved oxygen. The third level, the most complex, is quantitative description and prediction based on rate measurements and their relation to one another in terms of theoretical rate expressions. For our example, this would include such questions as how fast does sulfate reduction occur at each depth in the sediment and how does this rate depend on the concentrations of sulfate and of metabolizable organic matter. It is obvious that the study of diagenesis needs input at all three levels; in fact treatment at the third level is impossible without the gathering of information at the first two levels. Nevertheless, the writer feels that the third level has been, up to now, overly neglected.

The approach to diagenesis in this book is unorthodox in that it is entirely on the third level. This means that diagenesis is described quantitatively in terms of mathematical rate expressions. Our goal will be to develop theoretical models, based on a combination of sediment observations and *a priori* reasoning, which describe and interrelate in a quantitative manner the various diagenetic processes that affect a given sediment property during early diagenesis. Although many important diagenetic problems will be addressed, it is not our purpose to discuss each mineral type, sediment type, etc. in terms of natural observations and thermodynamic equilibrium (levels one and two). The topics covered in the last three chapters are chosen merely to illustrate the power of the mathematical models developed in the previous three chapters and not to provide a survey either of diagenetic processes or of the products of these processes. Thus, the scope of the book is narrow but very deep. This is done partly out of necessity because much of the rate data upon which diagenetic models must rest has not yet been obtained. For more comprehensive works concerned with diagenesis in general, as well as other aspects of sediments and sedimentary rocks (presented mainly on the first and second levels), the interested reader is referred to the books by von Engelhardt (1977); Berner (1971); Degens (1965); Blatt et al. (1972); Bathurst (1971); and Füchtbauer (1974).

By restricting ourselves to early diagenesis we unfortunately remove from discussion many interesting meteoric and deep-burial processes. However, because our approach is strictly theoretical, there are a number of practical advantages to studying only early diagenesis. For one thing, sediments undergoing early diagenesis can be sampled continuously and changes occurring both in solids and in pore waters can be examined in detail at all depths. This is accomplished by conventional coring or by shipboard drilling. Secondly, the sediment-water interface can be used as a reference plane for quantifying changes with depth, which in continually deposited sediments, also means time. This is an extremely important simplification when applying theoretical models to diagenetic problems. Third, because burial is to only a few hundred meters, temperatures do not appreciably exceed 25°C. This means that bacterial processes (which have an important effect on diagenesis) may occur at all depths because the temperatures are sufficiently low, that standard state thermodynamic data can be used as a first approximation to calculate chemical equilibria, and that complex physical processes accompanying heat flow over large depths may normally be ignored. Finally, the restriction to shallow depths means that large scale and deep burial phenomena, which are difficult to treat quantitatively and theoretically, can be ignored. This includes such complex processes as compaction by crushing and deformation of grains, salt filtering, and deep-seated ground water flow.

This book focuses mainly on the chemical composition of the interstitial waters of sediments. This is done for two reasons. First, pore waters are very sensitive indicators of incipient diagenetic changes in the solids. For example, a readily measurable 20% increase of dissolved Ca in a marine sediment pore water resulting from the dissolution of calcium carbonate is approximately equivalent, at depth in the sediment, to a decrease in the total solids of only 0.02% $CaCO_3$ by weight. If $CaCO_3$ is present as a major constituent (say 50%), this small relative change, because of historical fluctuations, cannot be detected by studying the solids alone. The second reason for emphasizing pore water chemistry is that pore waters, because they generally exhibit more striking changes during early diagenesis than the solids, have been studied much more (e.g., see Manheim, 1976). Also, chemical analyses of pore solutions are generally easier, and their interpretation is less ambiguous, than chemical analyses of the sediment solids. What we need at the present time are new and better ways of characterizing the chemical composition of solids (and their surfaces) in order to complement the many analyses of pore water chemistry that have already been made.

The approach of this book is mainly didactic. In Chapter 2 the basic theory used throughout the rest of the book is presented in terms of a general diagenetic equation. In Chapters 3, 4, and 5 detailed discussion of each of the terms in the general diagenetic equation is undertaken along with presentations of useful data. These chapters are intended to enable the reader to treat diagenetic problems in general in terms of rate models. In the last three chapters (6, 7, and 8), examples of diagenetic calculation are given using the theory developed earlier. These examples are taken from the recent literature and are divided into three categories: marine sediments of the continental margin, pelagic sediments, and non-marine sediments.

Part I

THEORY

2

General Theory

The purpose of this short chapter is to show, in the most general manner possible, how the subject of early diagenesis can be approached from a theoretical viewpoint. This serves both to set the stage for more detailed theoretical discussion of diagenetic processes in the succeeding three chapters, and to provide a fundamental basis for describing diagenesis throughout the remainder of the book. In order to understand diagenetic modeling properly, the reader should clearly understand the concepts presented here.

Preliminary Considerations

Let p represent any property of a sediment or sedimentary rock, for example, concentration of a certain mineral, grain size, faunal composition, or water content. Property p, at the time of deposition, will vary laterally due to variations in the environment of deposition. Upon burial, vertical variations in p will also arise either by historical changes at the site of deposition or by diagenesis. A complete description of p thus would entail a knowledge of both lateral and vertical variations and how they change with time. However, changes exhibited by sediments on the scale of a few hundred meters (as represented by early diagenesis) are predominantly vertical in nature (e.g., bedding, chemical gradients), and depth is uniquely different from the lateral dimensions in that it is directly related to geological time. Because of this, and because it is mathematically much simpler to deal with only one spatial dimension, we will treat our sediment property, p, in general, as though it were a function only of depth and time. (Lateral changes resulting from diagenesis, e.g., concretion formation, are discussed later as separate special topics.) Mathematically this can be expressed as:

$$p = f(x,t), \qquad (2\text{-}1)$$

where x = depth in the sediment, and
t = time.

For an ancient uplifted rock the origin for a depth coordinate would have to be fixed on some specific layer. By contrast, for sediments still undergoing deposition, we are presented with a choice of two origins for the depth coordinate. We may choose either a given layer (strictly speaking, an infinitely thin layer) or the sediment-water interface. Transformation from one origin to the other is expressed by:

$$\left(\frac{\partial p}{\partial t}\right)_{fixed\ depth} = \left(\frac{\partial p}{\partial t}\right)_{fixed\ layer} - \omega\left(\frac{\partial p}{\partial x}\right)_{fixed\ time}. \tag{2-2}$$

The parameter ω is most readily visualized as the rate of burial of the layer below the sediment-water interface, which, in the absence of complicating factors such as compaction and non-steady deposition, is simply the rate of deposition. Since in this book we will be concerned only with early diagenesis, we will adopt the sediment-water interface as origin. Thus, depth x will be measured positively downward from this interface. With this convention, and using a total derivative to denote changes "following the motion" of the layer, the above equation can be expressed in simplified notation as:

$$\left(\frac{\partial p}{\partial t}\right)_x = \frac{dp}{dt} - \omega\left(\frac{\partial p}{\partial x}\right)_t. \tag{2-3}$$

Mathematical representation of special sedimentary situations can be done by reference to equation (2-3), using the simplifying assumption that ω does not vary with time or depth (constant sedimentation rate with no compaction). The first situation, illustrated in Figure 2-1, is that of no diagenesis. This normally pertains to solids and means that all observed changes in p with depth are due to historical changes at the time of deposition. In other words, p within a given *layer* remains constant:

$$\frac{dp}{dt} = 0 \qquad \text{(no diagenesis)}, \tag{2-4}$$

and as a result,

$$\left(\frac{\partial p}{\partial t}\right)_x = -\omega\left(\frac{\partial p}{\partial x}\right)_t. \tag{2-5}$$

Since (by assumption) ω for a given layer does not vary with time, measurement of $\partial p/\partial x$ and the elapsed time represented by the thickness of the layer, enables us, via equation (2-5), to calculate the original rate of change of p at the time of sedimentation. This is the way the historical record is normally interpreted from sediments and sedimentary rocks. An example is when p represents the median size of sand grains which vary with depth due to original fluctuations in turbulence of water overlying the site of deposition. Without the assumption of no diagenesis, as represented by equation (2-4), much of historical geology would be impossible.

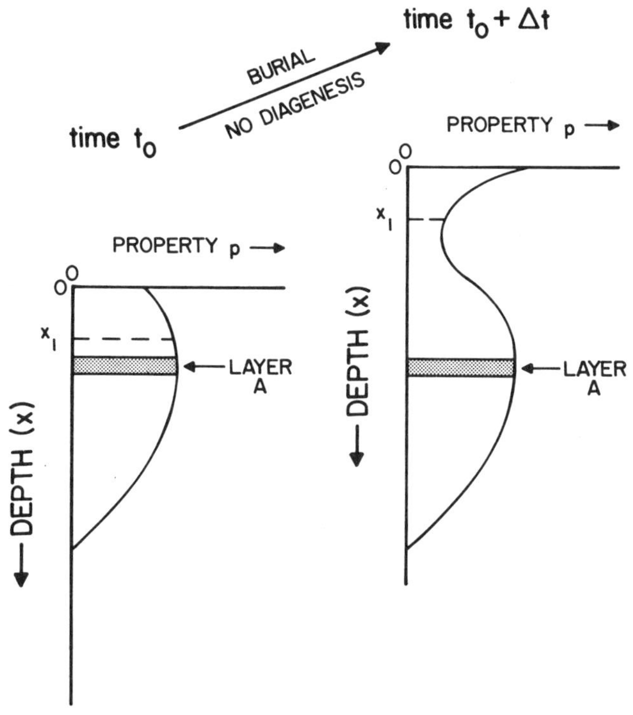

FIGURE 2-1. Diagrammatic illustration of situation of no diagenesis. Upon burial, the value of property p changes for a fixed depth, x_1 (or $x = 0$), but does not change for a given layer A.

The opposite situation, which is more important to the general theme of the present book, is where all observed changes in p with depth are due to diagenesis. This means that there have been no important historical fluctuations affecting p at the time of deposition, or, in other words, a layer at depth x, at the time it was deposited, had the same value of p as the layer currently being deposited. (Mathematically this corresponds to a constant upper boundary condition.) This gives rise to the situation of *steady state diagenesis* where p at a given *depth* (including the sediment-water interface) remains constant. Mathematically,

$$\left(\frac{\partial p}{\partial t}\right)_x = 0, \tag{2-6}$$

so that

$$\frac{dp}{dt} = \omega \left(\frac{\partial p}{\partial x}\right)_t. \tag{2-7}$$

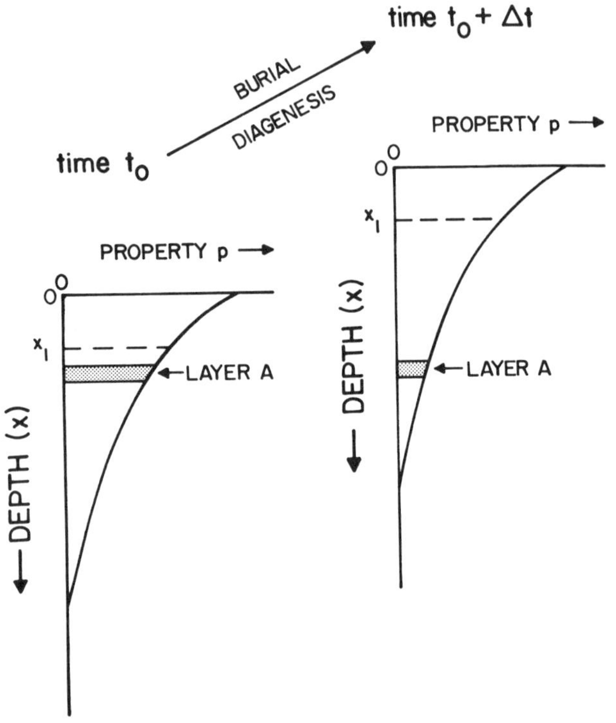

FIGURE 2-2. Diagrammatic illustration of situation of steady state diagenesis. Upon burial, the value of property p does not change for a fixed depth, x_1 (or $x = 0$), but does change for a given layer A. (Modified after Berner, 1971.)

This situation is illustrated in Figure 2-2. Note that the shape of the curve of p vs depth does not change during burial, while p for a given layer changes appreciably. The concept of steady state diagenesis is most useful, and will be employed throughout this book as an idealized model with which to compare real diagenetic situations. Also, steady state with respect to one property often entails steady state with respect to many others (although there is no *necessary* connotation).

A third situation is where there is steady state at all depths but no diagenesis. In other words,

$$\left(\frac{\partial p}{\partial t}\right)_x = 0, \tag{2-8}$$

$$\frac{dp}{dt} = 0, \tag{2-9}$$

so that

$$\left(\frac{\partial p}{\partial x}\right)_t = 0. \tag{2-10}$$

This results from either the presence of non-reactive p in the sediment or rapid attainment of equilibrium at the time of deposition, and is marked by a lack of change in p with depth.

So far, we have not discussed the kinds of sediment properties with which we will be principally concerned. Since the approach of this book is mainly chemical, the properties normally dealt with will be the concentrations of chemical components both in solids and in the interstitial water. However, because of their importance to compaction and bioturbation, we will also concern ourselves, but to a lesser extent, with the concentrations of total solids and total pore water.

The General Diagenetic Equation

Mass balance of material in a small box of sediment can be expressed simply: the difference between what comes in and what goes out and what is produced or consumed is equal to what accumulates in the box. Adopting the sediment-water interface as origin (i.e., a box of sediment between two fixed depths), this concept can be expressed mathematically as:

$$\frac{\partial \hat{C}_i}{\partial t} = -\frac{\partial F_i}{\partial x} + \sum R_i, \tag{2-11}$$

where $\hat{C}_i$ = concentration of solid or liquid component i in mass per unit volume of total sediment (solids plus pore water);

F_i = flux of component i in terms of mass per unit area of total sediment per unit time;

R_i = rate of each diagenetic chemical, biochemical, and radiogenic reaction affecting i, in terms of mass per unit volume of total sediment per unit time;

$\sum$ = summation sign.

(Note that here the subscript x is omitted from $\partial \hat{C}/\partial t$. To save printer's ink, subscripts will be omitted in future discussion, so that partial differentiations will always refer to fixed time and fixed depth below the sediment-water interface.)

Fluxes can be divided into two types, diffusive and advective. Diffusive fluxes arise statistically as the result of random motions of individual particles, ions, and the like, and they follow Fick's laws of diffusion (see next chapter under "Diffusion"). Advective fluxes, by contrast, are unidirectional due to an impressed force. Our major concern with advection will be with differential flux arising from compaction. However, at present we will not be concerned with the various types of diffusive and advective fluxes, but will merely express the two types in a generalized form. Thus:

$$F_i = -D\frac{\partial \hat{C}_i}{\partial x} + \upsilon\hat{C}_i, \tag{2-12}$$

where D = diffusion coefficient in area of total sediment per unit time;
υ = velocity of flow relative to the sediment-water interface.

For solids υ is simply equal to ω, the rate of burial as defined above. However, for pore water υ is in general different from ω, as explained in the next chapter. Combining equations (2-11) and (2-12),

$$\frac{\partial \hat{C}_i}{\partial t} = \frac{\partial\left(D\dfrac{\partial \hat{C}_i}{\partial x}\right)}{\partial x} - \frac{\partial(\upsilon\hat{C}_i)}{\partial x} + \sum R_i. \tag{2-13}$$

This expression shows that the three major processes affecting the concentration of a sedimentary component at a given depth are diffusion, advection, and diagenetic reaction. It is the most general expression for diagenesis that will be given in this book and, thereby, is designated as the *general diagenetic equation.*

If we prefer to express the general diagenetic equation in terms of a layer based coordinate system, we can transform (2-13) through the use of equation (2-3). The result is:

$$\frac{d\hat{C}_i}{dt} = \frac{\partial\left(D\dfrac{\partial \hat{C}_i}{\partial x}\right)}{\partial x} - \frac{\partial(\upsilon\hat{C}_i)}{\partial x} + \omega\frac{\partial \hat{C}_i}{\partial x} + \sum R_i, \tag{2-14}$$

which in the case of solid particles, where $\upsilon = \omega$, simplifies to

$$\frac{d\hat{C}_i}{dt} = \frac{\partial\left(D\dfrac{\partial \hat{C}_i}{\partial x}\right)}{\partial x} - \hat{C}_i\frac{\partial \omega}{\partial x} + \sum R_i. \tag{2-15}$$

This presentation allows us to turn, in the next three chapters, to detailed discussion of the diagenetic processes represented by each of the terms of these equations.

3

Diagenetic Physical and Biological Processes

In this chapter those processes which are included in the advective and diffusive terms of the general diagenetic equation are discussed in detail. Major subjects covered include: compaction, depositional burial, water flow, interstitial molecular diffusion, benthic boundary diffusion, bioturbation, and transfer across the sediment-water interface. Although considerable descriptive material is given, a mathematical description is attempted wherever possible. The major goal is to demonstrate how a wide variety of physical and biological processes can be described mathematically in terms of either advection or diffusion.

Advective Processes

Advection here refers to the bulk flow of solids or pore water relative to an adopted reference frame. If the sediment-water interface is used, advection of solids is due primarily to depositional burial. Burial rates of marine sediments range from about 0.1–10 cm per thousand years in deep-sea sediments to 0.1–10 cm per year in near-shore fine-grained muds. Episodic deposition, such as occurs in association with turbidity currents, can be considerably higher than average rates for any given environment. If there is negligible compaction or externally impressed flow (see below), advection of pore water is essentially the same as the rate of burial of solids.

In fine-grained sediments differential advection of particles and pore water normally occurs, and is brought about in most instances by compaction. Although the effects of compaction on diagenetic chemical changes can often be ignored in rough calculations (compactive flow past grains is always lower than particle burial rates), in principle it is important to be able to estimate these effects. In addition, compaction brings about appreciable changes in water content, which is a major property of sediments, and it can indirectly affect other diagenetic processes such as diffusion. For these reasons much of our discussion of advection will revolve around the subject of compaction.

Compaction

Compaction is here defined as the loss of water from a layer of sediment, due to compression arising from the deposition of overlying sediment. During early diagenesis this entails the closer packing together of solid particles with consequent upward expulsion of pore water (Burst, 1976) and is exhibited mainly by fine-grained muds. In order to simplify discussion of compaction, we introduce the parameter ϕ, or useful porosity (Engelhardt, 1977), which is defined as:

$$\phi = \frac{\text{volume of interconnected water}}{\text{volume of total sediment or rock}}$$

Although ϕ will be designated simply as porosity, strictly speaking it is not the same as total porosity. Total porosity refers to total void space per volume of sediment, whereas ϕ includes only connected pore space through which water may flow. In other words, isolated fluid-filled pores are excluded. Fortunately, these are rare in shallowly buried sediments, and as a result we will treat ϕ as if it were also total porosity. That is to say, isolated pores, if present, are included within the volume of total sediment solids. Also, we will assume that all pore space is filled with water, which obviously eliminates from discussion such subjects as unsaturated soils and petroleum-impregnated sands.

The effect of compaction upon sediment advection can best be seen in terms of general diagenetic equations for total solids and water in pore solution. From (2-13) they are:

$$\frac{\partial \hat{C}_{ts}}{\partial t} = \frac{\partial\left(D_{ts}\dfrac{\partial \hat{C}_{ts}}{\partial x}\right)}{\partial x} - \frac{\partial(\hat{C}_{ts}\omega)}{\partial x} + R_{ts}, \tag{3-1}$$

$$\frac{\partial \hat{C}_{w}}{\partial t} = \frac{\partial\left(D_{w}\dfrac{\partial \hat{C}_{w}}{\partial x}\right)}{\partial x} - \frac{\partial(\hat{C}_{w}v)}{\partial x} + R_{w}, \tag{3-2}$$

where ω = velocity of burial of solid particles below the sediment-water interface;

v = velocity of water flow relative to the sediment-water interface;

ts = total solids;

w = water in pore solution.

Now, we can recast the concentration $\hat{C}$ in terms of more commonly used parameters:

$$\hat{C}_{ts} = \bar{\rho}_s(1 - \phi), \tag{3-3}$$

$$\hat{C}_{w} = \rho_w^*\phi, \tag{3-4}$$

where $\bar{\rho}_s$ = average density of solids;

ρ_w^* = mass of water per unit volume of pore solution (not density of the solution).

For the purposes of the present chapter we will not be concerned with diffusion of total solids and pore solution which arise mainly as a result of bioturbation. (For a recent discussion of the effects of bioturbation upon porosity consult Hakanson and Kallstrom, 1978.) Therefore, to simplify discussion, the first terms on the right-hand side in equations (3-1) and (3-2) will be dropped. Then, upon substitution of (3-3) and (3-4) in (3-1) and (3-2) respectively:

$$\frac{\partial[\bar{\rho}_s(1-\phi)]}{\partial t} = -\frac{\partial[\bar{\rho}_s(1-\phi)\omega]}{\partial x} + R_{ts}, \tag{3-5}$$

$$\frac{\partial(\rho_w^*\phi)}{\partial t} = -\frac{\partial(\rho_w^*\phi v)}{\partial x} + R_w. \tag{3-6}$$

Two further simplifications can be made. Chemical reactions involving total solids or pore water are in most cases unimportant. (Two major exceptions are cementation and the dissolution of $CaCO_3$ in calcareous-rich deep-sea sediments, which are discussed later as separate problems.) Thus, R_{ts} and R_w we will assume to be zero. In addition, density changes of total solids $\bar{\rho}_s$ and mass of water per unit volume of pore solution ρ_w^* with depth are small and will also be ignored. With these simplifications equations (3-5) and (3-6) reduce to:

$$\frac{\partial(1-\phi)}{\partial t} = -\frac{\partial[(1-\phi)\omega]}{\partial x}, \tag{3-7}$$

$$\frac{\partial\phi}{\partial t} = -\frac{\partial(\phi v)}{\partial x}, \tag{3-8}$$

and thus,

$$\frac{\partial[(1-\phi)\omega]}{\partial x} = -\frac{\partial(\phi v)}{\partial x}. \tag{3-9}$$

(Equation (3-9) does not imply that $\partial(1-\phi)\omega/\partial t = -\partial(\phi v)/\partial t$). These are the expressions normally used to express the effects of compaction on the advection of solids and pore water (Smith, 1971; Berner, 1975; Schink and Guinasso, 1978a). Note that they are strictly true only in the absence of bioturbational or other mixing. In terms of a sediment layer they can be transformed via equation (2-3) to:

$$\frac{d\phi}{dt} = -\frac{\partial(\phi v)}{\partial x} + \omega\frac{\partial\phi}{\partial x} = (1-\phi)\frac{\partial\omega}{\partial x}. \tag{3-10}$$

Introducing the parameter $v_g \equiv v - \omega$, which is the velocity of pore water relative to the layer, we have:

$$\frac{d\phi}{dt} = -\frac{\partial(\phi v_g)}{\partial x} - \phi\frac{\partial\omega}{\partial x}. \tag{3-11}$$

Examination of these porosity equations reveals some important relationships regarding advection and compaction. First, consider the situation of no compaction, which would be a good approximation for surficial sands. If there is no compaction, porosity within a given layer does not change during burial; in other words, $d\phi/dt = 0$. If so, we obtain from

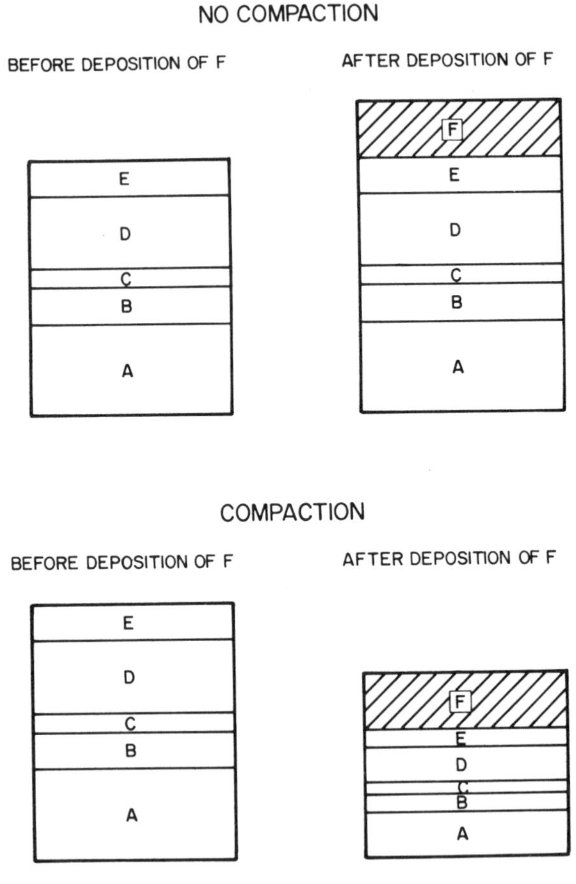

FIGURE 3-1. Comparison of burial with and without compaction. Here letters A–F refer to sediment layers. In the case of no compaction, all layers and horizons are buried at the same rate equal to the thickness of the layer (F) being currently deposited at the sediment-water interface. Also, none of the layers undergo thinning. With compaction, layers are thinned and do not undergo burial at the same rate. For instance, the base of layer A actually approaches the sediment-water interface, due to compaction, while the interface between layers E and F undergoes burial equal (as before) to the thickness of layer F.

equations (3-7), (3-10), and (3-11):

$$\partial\omega/\partial x = 0 \qquad \text{(no compaction),} \qquad (3\text{-}12)$$

$$\frac{\partial\phi}{\partial t} = -\omega\frac{\partial\phi}{\partial x} \qquad \text{(no compaction),} \qquad (3\text{-}13)$$

$$\frac{\partial(\phi v_g)}{\partial x} = 0 \qquad \text{(no compaction).} \qquad (3\text{-}14)$$

Equation (3-12) states that, without compaction, the sediment is buried as a solid body at the rate at which new material is added at the sediment-water interface. Comparison of burial under the situations of compaction and no compaction is illustrated in Figure 3-1. Equation (3-13) shows that, even if there is no compaction, there can still be variations of porosity at a given *depth*. This would be due to the successive burial of layers with different initial porosities. Equation (3-14) states that, in the absence of compaction, flow through a sediment layer is incompressible. This expression is the one usually used by ground water hydrologists to describe flow through water-saturated, indurated rocks (e.g., Bear, 1972; Domenico, 1972).

The assumption of no compaction also allows further simplification. One can consider the velocity, v, of pore water to be made up of two components, one due to compaction, and the other to externally impressed flow (e.g., Imboden, 1975). Externally impressed flow is what is normally thought of when one discusses ground water flow in sedimentary rocks. An example in unlithified subaqueous sediments would be the submarine discharge of fresh water percolating up through the sediments. If there is no externally impressed flow, which is a much more common situation in modern sediments, then v_g results only from compaction. If, in addition, there is no compaction, then $v_g = 0$ and

$$v = \omega$$

In other words, in the absence of compaction and externally impressed flow, both solids and included pore water are buried at the same rate. This simple situation will often be used in this book when constructing diagenetic models.

If compaction is non-negligible (and externally impressed flow can be ignored) a useful concept is that of steady state compaction. This means that the porosity changes only as a result of compaction. In other words, sedimentation rate is constant and $\partial\phi/\partial t = 0$ at all depths. This would be appropriate for the common early-diagenetic situation of continuous deposition of uniform fine-grained muds. If there is steady state compaction, then from (3-7) and (3-8):

$$\frac{\partial[(1-\phi)\omega]}{\partial x} = 0 \qquad \text{(steady state compaction),} \qquad (3\text{-}15)$$

$$\frac{\partial(\phi v)}{\partial x} = 0 \qquad \text{(steady state compaction).} \qquad (3\text{-}16)$$

Equation (3-15) shows that, as ϕ decreases with depth due to compaction, ω also decreases. This is illustrated in Figure 3-2. If compaction and, thus,

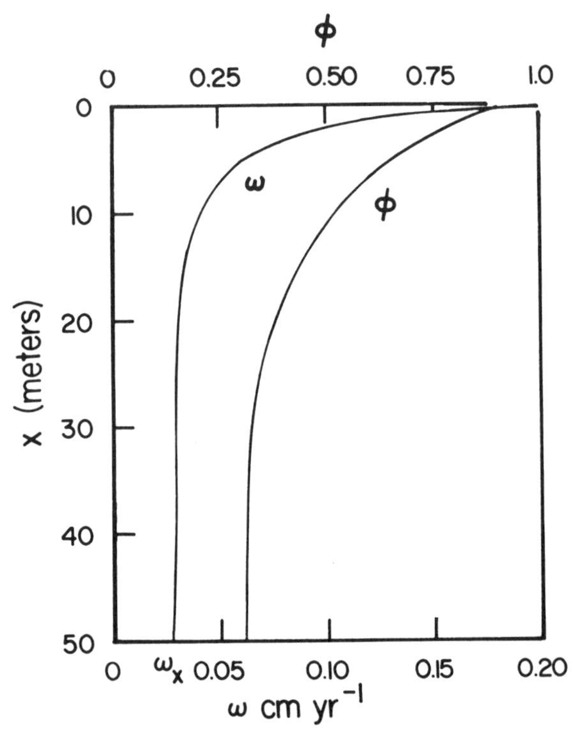

FIGURE 3-2. Effect of decreasing ϕ, due to steady state compaction, on ω, the rate of burial of solids. A hypothetical porosity distribution has been assumed to follow the relation:

$$\phi = 0.60 \exp(-0.10x) + 0.30,$$

where x is depth in meters. At steady state compaction using equation (3-15) this leads to the relation:

$$\omega = \frac{\omega_0}{[7 - 6\exp(-0.10x)]},$$

where ω_0 = rate of burial at $x = 0$. Here the value, $\omega_0 = 0.20$ cm yr^{-1} is assumed. Note the asymptotic value, $\omega = \omega_x$.

the porosity gradient become very small ($\partial\phi/\partial x \to 0$), below a certain depth x:

$$(1 - \phi)\omega = (1 - \phi_x)\omega_x, \tag{3-17}$$

$$\phi v = \phi_x v_x. \tag{3-18}$$

Since with negligible compaction $v_x = \omega_x$, equations (3-17) and (3-18) can be rewritten as:

$$\omega = \frac{(1 - \phi_x)}{(1 - \phi)}\,\omega_x, \tag{3-19}$$

$$v = \frac{\phi_x}{\phi}\,\omega_x. \tag{3-20}$$

Thus, if porosity is known as a function of depth, and the burial rate at depth where compaction is negligible ω_x is also known, one can calculate ω or v at any other depth if one assumes steady state compaction. This is also shown in Figure 3-2. (Strictly speaking, porosity tends to diminish continuously until very great depths are reached. However, for the uppermost few one hundred meters of sediment, porosity profiles often show an *essentially* asymptotic character so that the concept of ω_x is quite useful.)

Rate of Deposition

Throughout our discussion we have implicitly assumed that ω, the rate of burial of solids, is somehow equivalent to the "rate of deposition." This is strictly true only if sedimentation rate is constant and compaction is absent or at steady state. Here we address ourselves to problems presented by non-constant depositional rate. Non-constant deposition means that ω changes with time, i.e., $\partial\omega/\partial t \neq 0$. In this case the rate of deposition of a given layer, as deduced from dating of its upper and lower boundaries, is not the same as ω. This can be seen by reference to Figure 3-3. The rate of

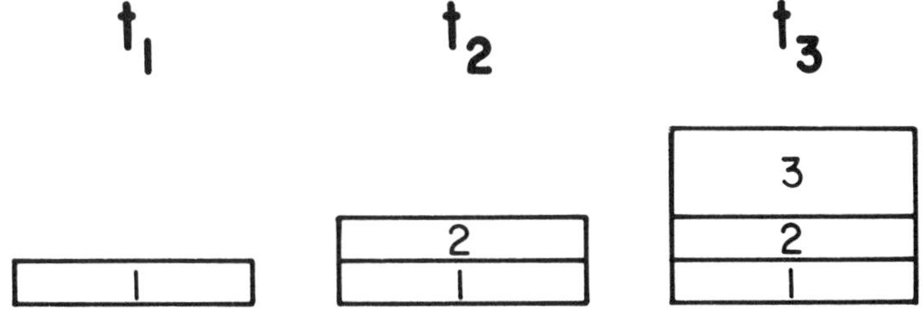

FIGURE 3-3. Non-steady state deposition (without compaction). Each layer, marked 1, 2, and 3, represents the same interval of time. The *rate of burial* $\omega(0,t)$ of layer 1 during time $(t_3 - t_2)$ is equal to the rate of deposition of layer 3, which is *twice the original rate of deposition* $\omega(0,0)$ of layer 1. (Symbols t_1, t_2, and t_3 refer to times after deposition of layers 1, 2, and 3 respectively.)

deposition of layer 1 is its thickness divided by the time interval, $t_1 - t_0$. During an equivalent interval $t_3 - t_2$, twice as much sediment is deposited (layer 3), and as a result, layer 1 is buried twice as fast. Thus, at time t_3, the burial rate of layer 1 is twice its measured sedimentation rate. In other words (for no compaction):

$$\omega(x,t) = \omega(0,t) \neq \omega(0,0), \tag{3-21}$$

where $\omega(0,0)$ represents original burial rate which is equivalent to the measured rate of deposition.

If we are dealing with ancient rocks, generally $\omega(x,t)$ is a non-determinable and useless parameter. All we can determine is $\omega(0,0)$ (once proper correction for compaction is made). However, in modern sediments still undergoing sedimentation, it behooves us to try to deduce $\omega(x,t)$. If there is no compaction, this is done simply by measuring $\omega(0,t)$, the burial rate of newly added sediment. By contrast, if there is compaction, the problem becomes much more difficult.

The relation between $\omega(x,t)$ and measured rate of sedimentation, for non-constant deposition with compaction, is discussed by Imboden (1975), who derives the expression:

$$\omega(x,t) = s(0,t) + \int_0^x \frac{1}{s}\frac{ds}{dt}\,dx, \tag{3-22}$$

where $s(0,t) = \omega(0,t)$;

s = "rate of deposition" or the measured thickness representing a specified unit of time (more simply one can think of it as the thickness of an annual layer);

X = depth of layer of interest;

and the total derivative (ds/dt) refers to each layer. The problem with equation (3-22), is that (ds/dt) for each layer can be known only from measurements of the same sediment taken at (at least) two times that are significantly different on a sedimentological time scale. Such measurements are usually impossible. Thus, various additional assumptions need to be made, the most simple of which is constant sedimentation rate combined with steady state compaction. Under these conditions (Imboden, 1975):

$$\omega(x,t) = s(x,t), \tag{3-23}$$

and the problem reduces simply to correcting ω (or s) for steady state compaction. The less restrictive assumption of steady state compaction with non-constant sedimentation is rather untenable. This is because more rapid deposition generally leads to a higher initial porosity even for sediments with the same mineralogy, compaction, etc. (see under "Factors affecting compaction and porosity" below). Thus, non-constant sedimentation implies non-steady state compaction, and the two cannot be separated.

In order to avoid unnecessary complexity, we will assume in all future discussions that constant sedimentation rate and steady state compaction exist, and thus, that ω is equivalent to the measured rate of sedimentation. Whenever ω is used, these assumption should always be kept in mind.

The phrase, "rate of sedimentation" is often thought of in terms of a flux of sediment particles to the sediment surface and not as dx/dt as we have used it. Conversion between sediment flux rate and ω is simple if we view ω as volume of sediment added to each unit area of sediment surface per unit time. Then:

$$\mathcal{R} = \bar{\rho}_s\omega(1 - \phi), \tag{3-24}$$

where $\mathcal{R}$ = rate of sediment flux in mass per unit area per unit time (assumed constant with depth and time);

$\bar{\rho}_s$ = average density of solids.

Note that this relation holds for each depth, x, whether or not there is compaction.

Some Additional Useful Relations

Assumption of steady state compaction allows the calculation of some useful parameters from plots of porosity vs depth. For these calculations changes of porosity due to bioturbation and chemical reaction are ignored. The parameters are:

1. *Age of a layer.* This is the elapsed time, τ, between the present and the time the layer was originally deposited. Under steady state compaction, τ is given by:

$$\tau = \int_0^{x'} \omega^{-1}\, dx, \tag{3-25}$$

where x' = present depth of the layer. Substituting equation (3-19) in (3-25):

$$\tau = \frac{1}{(1 - \phi_x)\omega_x} \int_0^{x'} (1 - \phi)\, dx, \tag{3-26}$$

where ϕ_x and ω_x refer to values at depth where $\partial\phi/\partial x \approx 0$.

2. *Rate of compaction within a layer.* This is represented by $d\phi/dt$. Using the coordinate transformation equation (2-3):

$$\frac{\partial\phi}{\partial t} = \frac{d\phi}{dt} - \omega\frac{\partial\phi}{\partial x}. \tag{3-27}$$

At steady state compaction, $\partial\phi/\partial t = 0$, so that:

$$\frac{d\phi}{dt} = \omega \frac{\partial\phi}{\partial x}. \tag{3-28}$$

3. *Total compaction of a layer since deposition.* This is calculated from the fact that the volume of solids (per unit area) within a layer does not change during compaction. Mathematically:

$$(1 - \phi)h_x = (1 - \phi_0)h_0, \tag{3-29}$$

where h_x = thickness of a layer at depth x;

h_0, ϕ_0 = original thickness and porosity at the time of deposition of the layer.

Since there is steady state compaction, h_0 and ϕ_0 also refer to the layer currently being deposited. Rearranging equation (3-29) we obtain the degree of compaction $(h_0 - h_x)/h_0$:

$$\frac{h_0 - h_x}{h_0} = \frac{\phi_0 - \phi}{1 - \phi}. \tag{3-30}$$

4. *Rate of water flow through a layer due to compaction.* This is represented by $Q_g = \phi v_g$; v_g represents compactive flow only. By definition:

$$\phi v_g = \phi v - \phi\omega. \tag{3-31}$$

Therefore, from equation (3-20) for steady state compaction:

$$Q_g = \phi_x\omega_x - \phi\omega. \tag{3-32}$$

Note that, since $\phi\omega > \phi_x\omega_x$, $Q < 0$. In other words, compactive flow is upward relative to the layer.

5. *Rate of water flow through the sediment-water interface.* This is represented by $Q_0 = \phi_0 v_0$ and, again, we will ignore externally impressed flow. From equation (3-20):

$$Q_0 = \phi_x\omega_x. \tag{3-33}$$

Since for constant deposition $\omega_x > 0$, then $Q_0 > 0$. This means that in the presence of steady state compaction, water flow occurs *into* the sediment and not *out of* it as is often stated in the literature. Compactive flow of water out of a sediment can occur only under non-steady state conditions, such as rapid deposition of high porosity sediment followed by settling and compactive expulsion of the water.

6. *Total volume of water which has passed through a layer (horizon) since deposition.* This is represented by W_T (volume per unit area of sediment).

Under steady state compaction this is equal to the total volume of water in the whole sediment column at the time of deposition of the layer ($t = 0$) minus the volume of water below the layer at present ($t = \tau$). (See Figure 3-4.) Mathematically:

$$W_T = \left[\int_0^{x_B} \phi\, dx\right]_{t=0} - \left[\int_{x_P}^{x'_B} \phi\, dx\right]_{t=\tau}, \tag{3-34}$$

where x_P = present depth of the layer;

x_B, x'_B = depths at time 0, and time τ, respectively, where continuous upward flow is interrupted due to encountering impermeable basement or sand layers which act as lateral conduits.

Now, assuming an exponential-like porosity-vs-depth distribution, we have the relation:

$$x'_B = x_B + \omega_x \tau, \tag{3-35}$$

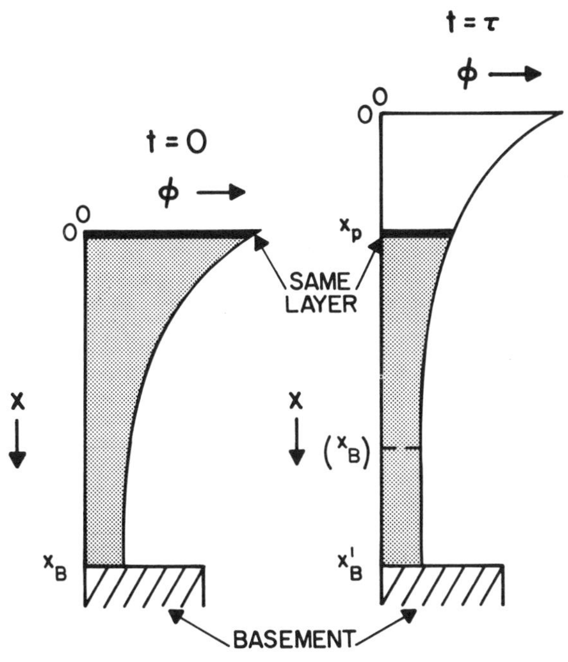

FIGURE 3-4. Total amount of water that has passed through a layer (strictly speaking an horizon) since original deposition for the situation of steady state compaction. This is equal to the integrated area below the sediment-water interface at the time of deposition ($t = 0$) (shown in gray on left) minus the integrated area currently below the layer (shown in gray on right).

so that

$$\left[\int_0^{x'_B} \phi\, dx\right]_{t=\tau} = \left[\int_0^{x_B} \phi\, dx\right]_{t=0} + \phi_x \omega_x \tau. \tag{3-36}$$

Substituting (3-26) and (3-35) in (3-34):

$$W_T = \int_0^{x_P} \phi\, dx - \frac{\phi_x}{1 - \phi_x} \int_0^{x_P} (1 - \phi)\, dx. \tag{3-37}$$

Factors Affecting Porosity and Compaction

The initial porosity of a sediment at the time of sedimentation is primarily a function of grain size. Small grains have high specific surface area and, consequently, are considerably influenced by surface chemical-electrostatic effects. This is best shown by the clay minerals which are present as small platelets of micron-to-submicron diameter. Electrostatic repulsion by like-charged basal planes, and attraction by oppositely charged edge and basal planes results in an open structure resembling a house of cards (Rieke and Chilingarian, 1974; Engelhardt, 1977). This structure is exhibited by both bulk sediment and individually sedimenting floccules, and results in very high initial porosities ($\phi = 0.7$ to 0.9). This is shown in Figure 3-5. Initial packing must also be affected by the properties of organic molecules which invariably coat the surfaces of clays (Rhoads, 1974), as well as by the nature and type of clay mineral. Solution

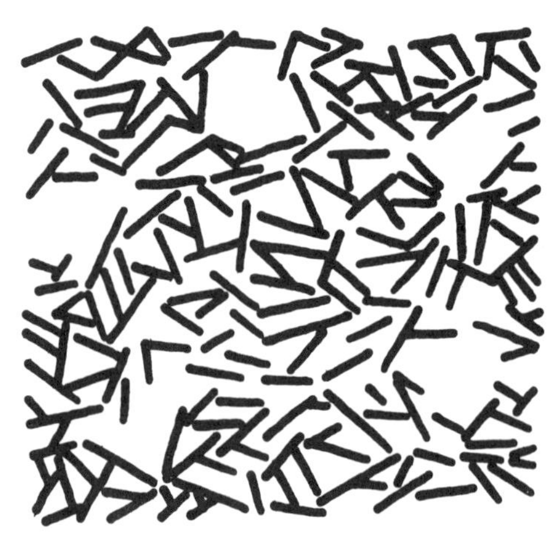

FIGURE 3-5. Idealized "house of cards" structure of flocculated clay. The individually sedimenting floccules upon touching one another produce the larger inter-floccule pores. (After Engelhardt and Gaida 1963.)

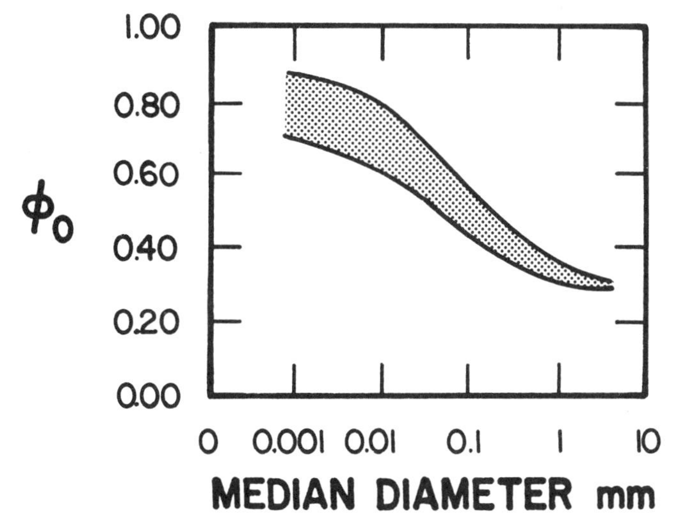

FIGURE 3-6. Initial porosity ϕ_0 as a function of grain size for terrigenous surfical sediments. The increase of porosity with decreasing median size reflects increasing proportions of clay minerals. (After Meade, 1966.)

composition and ionic strength also have an effect, as they control the nature and thickness of the electrical double layer (see under "Equilibrium absorption" in Chapter 4) but, unfortunately, no simple generalization concerning packing and solution composition can be made (Meade, 1966).

The initial porosity of coarse, sand-sized particles is much less affected by surface chemistry. Instead, simple geometric packing takes place which results in much lower porosities. For uniform spheres (regardless of size) one may calculate porosities ranging from $\phi = 0.26$ to 0.48 depending upon the closeness of the packing. Actual well-sorted sands exhibit measured porosities ranging from 0.36 to 0.46 (Engelhardt, 1977). The natural sands do not attain the lower calculated porosity, 0.26 which corresponds to hexagonal closest packing, because of resistance to re-orientation by the slightly angular grains. Although angularity is significant, the most important factor affecting the initial porosity of sand is sorting. Poor sorting enables interstices between larger grains to be filled by smaller grains, thereby lowering the porosity. The combined effects of all factors upon the initial porosities of natural sands, silts, and clays is shown, as a function of median particle size, in Figure 3-6.

Another factor affecting initial porosity of fine-grained sediments is bioturbation. Construction of burrows, and constant irrigation of these burrows results in a higher water content of sediments than would result in the absence of bioturbation (Rhoads, 1974; Hakanson and Kallstrom, 1978). Also, constant ingestion and defecation of fine particles by deposit-feeding organisms affects the organic content of the particle surfaces, and as a result, changes in surface chemistry must take place. The effect of bioturbation on both initial porosity and initial clay platelet orientation deserves considerably more attention than it has received in the past.

Because of differences in initial packing, the behavior during compaction of sands and clays is decidedly different. In the upper few hundred meters sands undergo only minor particle re-orientation and, as a result, decrease of porosity with depth is minimal. For the purposes of this book, compaction of sands will be neglected. By contrast, fine-grained clay-rich sediments undergo continual compaction (Figure 3-7), even on a centimeter-by-centimeter basis. The weight of overlying sediment begins to force mutually repelling particles closer together and to collapse the house of cards. Engelhardt (1977) describes this process in terms of an increase in grain-to-grain contacts (increasing coordination number) and the loss of pore space between originally deposited floccules (see Figure 3-5).

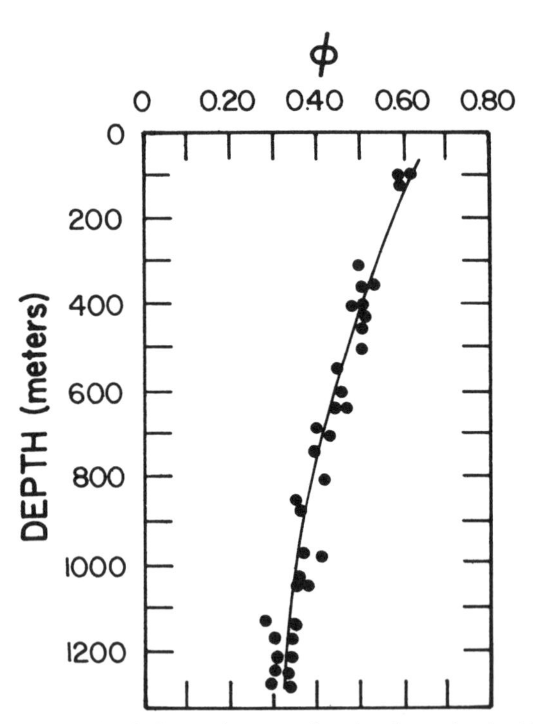

FIGURE 3-7. Porosity vs depth for terrigenous silty-clay from the Arabian Sea. (After Hamilton, 1976.)

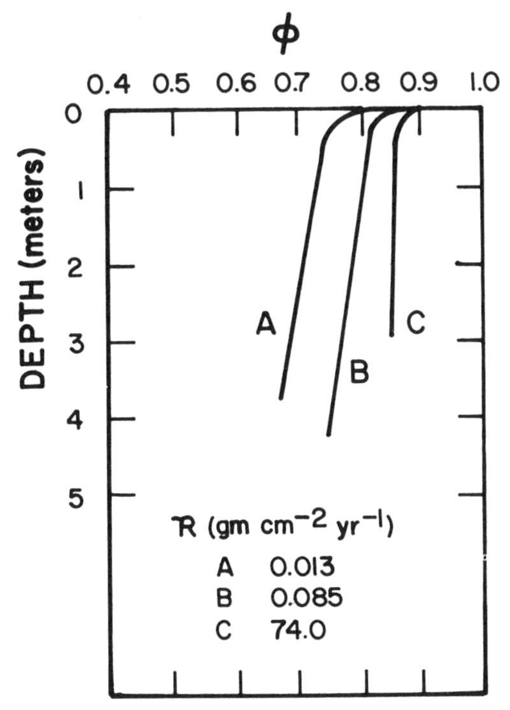

FIGURE 3-8. Porosity ϕ vs depth for fine-grained clayey sediments. Curve *C* refers to Lake Mead; curves *A* and *B* to basins off southern California. $\mathscr{R}$ = rate of deposition. (Adapted from Emery, 1960.)

Compaction of clay-rich sediments in the top few meters is shown in Figure 3-8. Differences in water content can be explained by possible differences in clay mineralogy, but also by differences in rate of sedimentation. Adjustment of porosity to an applied overburden is not instantaneous because water must be expelled. Thus, rapidly deposited sediment has had less time to adjust to the overburden and as a result, its water content is higher. In fact, water content (once the effects of bioturbation are taken into consideration) is often a good qualitative indication of deposition rate for surficial muds.

During burial below the top few meters clay-rich sediments undergo continual compaction, but the rate of compaction decreases with depth. This is evidenced (see Figure 3-9) in many sediments by a sharp drop in $\partial\phi/\partial x$ in the top few hundred meters, and can be viewed as a decrease in compressibility of the sediment with depth. (For a detailed discussion of compressibility consult Rieke and Chilingarian, 1974). Decrease in compressibility with increase in compaction is brought about by an increase

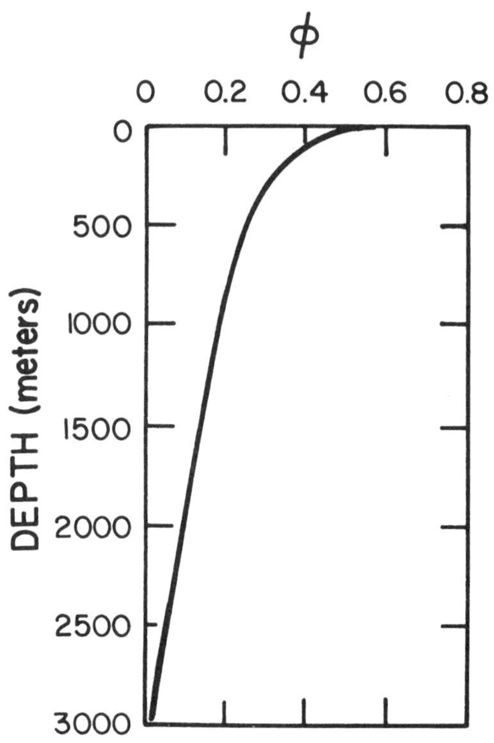

FIGURE 3-9. Porosity of clay sediments. Composite based on data for Recent, Tertiary, and Lias sediments. (After Engelhardt, 1977.)

in the number of stable contacts between grains (Engelhardt, 1977), which results in greater resistance to further deformation. As a corollary to this, compressibility from one sediment to another correlates well with initial porosity, with those sediments with the highest initial porosity being most compressible.

At depths of a few hundred meters where appreciable compressibility decreases are encountered, the resistance to vertical pore water flow begins to become important (Burst, 1976). This leads at a greater depth to the expulsion of pore water mainly in a *lateral* direction where permeability is higher. Dewatering takes place via lateral migration to permeable conduits such as sand bodies or fracture zones. Thus, the models presented in this section which treat compactive water flow as being only vertical in nature, begin to break down at depths greater than 500 meters. Also, upon very deep burial, both clays and sands undergo compaction not by particle reorientation but by the deformation, breakage, interpenetration, and cementation of grains. Such processes as lateral flow and particle deformation are important to the process of lithification but are beyond the scope of the present book.

Diffusion

Diffusion is the net motion of matter resulting from the random motion of individual entities, be they ions, sand grains, or drunken sailors. Diffusion in sediments can be divided into four categories: molecular diffusion (including surface diffusion), dispersion, wave and current stirring, and bioturbation (when treated as a diffusive process). Molecular diffusion, as the name implies, refers to the diffusion within a single phase of its atomic constituents, e.g., atoms, ions, and molecules. Dispersion (Bear, 1972) is enhanced diffusion of dissolved species in flowing groundwater as a result of local variations in the velocity along tortuous flow paths. It increases in importance as average flow rate increases. Wave and current stirring is a surficial process affecting only newly deposited sediment. Bioturbation is the mixing of sediment by organisms, which is often modeled as a diffusive process. Because of its unique character as well as its importance to the study of near-surface sediments, bioturbation is discussed here in a separate section.

In this book, theoretical discussion of flow in sediments is restricted to that arising from compaction. In other words, externally impressed ground water flow is ignored, because it is rare, or a poorly understood phenomenon in subaqueous, near-surface sediments. Also, flow due to compaction in fine-grained muds is very slow and in sands is non-existent. Thus, in either case dispersion due to flow can be ignored (Bear, 1972) and we can assume that, once depths below the zones of bioturbation and wave and current stirring are reached, the only diffusive process of any importance is molecular diffusion.

Molecular Diffusion

Molecular diffusion in a sediment can occur within the solids, on the "surface" of the solids, or in the interstitial water. Diffusion within solids at sedimentary temperatures is negligible, except in very small particles over long geological time scales. It will be ignored here. Surface diffusion is important only at low salinity, and/or high degrees of compaction accompanying deep burial. At salinities and porosities of marine sediments buried up to 500 m, surface diffusion should be negligible compared to pore water diffusion (Turk, 1976; McDuff and Gieskes, 1976). Thus, only pore water molecular diffusion will be discussed here and, to avoid undue wordiness the term "diffusion" will be used to refer to this process.

Diffusion in aqueous solution takes place in accordance with Fick's laws of diffusion (Crank, 1975) which, for uncharged species, are:

$$\textit{First Law:} \qquad J_i = -D_i \frac{\partial C_i}{\partial x}. \tag{3-38}$$

$$\textit{Second Law:} \qquad \frac{\partial C_i}{\partial t} = -\frac{\partial J_i}{\partial x} = \frac{\partial \left(D \dfrac{\partial C_i}{\partial x} \right)}{\partial x}. \tag{3-39}$$

$$\text{For constant } D\text{:} \quad \frac{\partial C_i}{\partial t} = D_i \frac{\partial^2 C_i}{\partial x^2}. \tag{3-40}$$

J_i = diffusion flux of component i in mass per unit area per unit time;

C_i = concentration of component i in mass per unit volume;

D_i = diffusion coefficient of i in area per unit time;

x = direction of maximum concentration gradient.

A diagrammatic derivation of the Second Law (with constant D) is shown in Figure 3-10. In the general diagenetic equation, these one-dimensional forms, modified for sediments, are what is used with x representing depth. However, in some special situations, such as concretion growth, other coordinate systems must be used. Fick's Second Law for spherical symmetry is:

$$\frac{\partial C_i}{\partial t} = \frac{1}{r^2} \frac{\partial \left(r^2 D \dfrac{\partial C_i}{\partial r} \right)}{\partial r}, \tag{3-41}$$

where r = (spherical) radial coordinate; and for cylindrical symmetry is:

$$\frac{\partial C_i}{\partial t} = \frac{1}{r} \frac{\partial \left(r D \dfrac{\partial C_i}{\partial r} \right)}{\partial r}, \tag{3-42}$$

where r = (cylindrical) radial coordinate.

Before applying Fick's laws directly to sediments, several modifications must be made. The first is correction for electrical effects. The major species that undergo pore water diffusion are ions. An ion migrates in response not only to its own concentration gradient (Fick's First Law) but also to gradients in electrical potential brought about by concentration gradients of the other ions (e.g., Anderson and Graf, 1978). Natural solutions contain many different ions, so that these electrical effects can be appreciable.

FICK'S SECOND LAW

CONSIDER RECTANGULAR BOX:

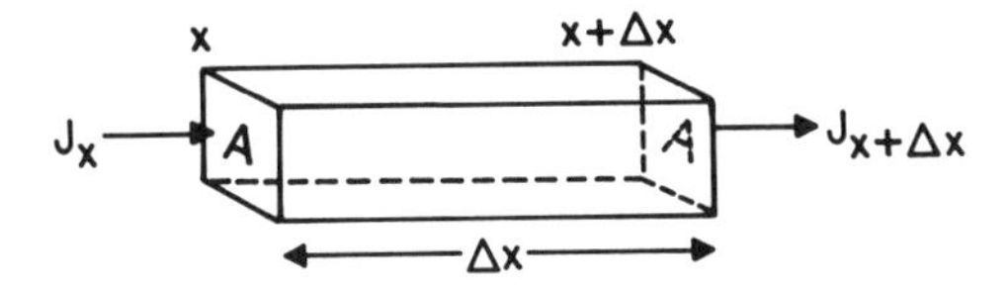

IF M = MASS:

$$\frac{\Delta M}{\Delta t} = -A(J_{x+\Delta x} - J_x)$$

FOR CONCENTRATION C:

$$\Delta C = \Delta M / A\Delta x$$

SUBSTITUTING WITH FICK'S FIRST LAW:

$$\frac{\Delta C}{\Delta t} = \frac{D\left[(\partial C/\partial x)_{x+\Delta x} - (\partial C/\partial x)_x\right]}{\Delta x}$$

PASSING TO LIMIT Δx → 0:

$$\frac{\partial C}{\partial t} = D\frac{\partial^2 C}{\partial x^2}$$

FIGURE 3-10. Derivation of Fick's Second Law of Diffusion.

Correction for electrical effects here follows the treatment of Vinograd and McBain (1941), and Ben-Yaakov (1972) as modified by McDuff and Ellis (1979). Ben-Yaakov gives the expression:

$$J_i = -m_i\left(RT\frac{\partial C_i}{\partial x} + Z_iC_iF\frac{\partial E}{\partial x}\right), \tag{3-43}$$

where m_i = mobility of ion i at infinite dilution;
E = electrical potential;
F = Faraday constant;
Z_i = valence of i;
T = absolute temperature;
R = gas constant.

This equation assumes that variation of the activity coefficient along the diffusion path is small. Since ion flux cannot result in the pile-up of charge, one also has the electroneutrality relation:

$$\sum ZJ = 0, \tag{3-44}$$

where summation is over all ions. Substitution of equation (3-43) for each diffusing species into (3-44) results in:

$$\frac{\partial E}{\partial x} = -RT\frac{\sum[Zm(\partial C/\partial x)]}{F\sum Z^2 mC} \tag{3-45}$$

If we now apply Fick's First Law to ion i,

$$J_i = -D_i\frac{\partial C_i}{\partial x}, \tag{3-46}$$

by combining equations (3.43), (3-45), and (3-46) we obtain (McDuff and Ellis, 1979):

$$D_i = m_i RT\left\{1 - \frac{Z_i C_i \sum[Zm(\partial C/\partial x)/(\partial C_i/\partial x)]}{\sum[Z^2 m^2 C]}\right\}. \tag{3-47}$$

In this form one can see that the Fick's Law diffusion coefficient, under the assumption of constant activity coefficients, is made up of two terms, one depending on mobility (at infinite dilution) and temperature, and the other on electrical effects due to the other ions.

An additional inter-ion effect is brought about by ion-pairing (Lasaga, 1979). When we speak of an ion diffusing in a pore solution, we generally refer to the sum of all major species containing the element of interest. For example, for sulfate "ion" in sea water this would include the free SO_4^{--} ion as well as the ion-pairs $MgSO_4^0$, $CaSO_4^0$, $NaSO_4^-$, and KSO_4^-. Yet equation (3-47) refers only to each distinct species. By assuming local equilibrium between all free ions and ion pairs Lasaga (1979) has included ion pairing in the effects of other ions on diffusion coefficients to give an expression more complicated than equation (3-47) above. His results for the major ions of seawater are given in Table 3-1.

Table 3-1 shows that for most ions (e.g., sulfate) the effects of ion pairing and electrical interactions on diffusion are small. However, for certain ions they can be appreciable. To gain an idea of the maximum effect of electrical coupling consider the diffusion of total Mg^{++}. Under extreme conditions of sulfate reduction, concentration gradients of sulfate in sediments may reach $\partial C_i/\partial x = 1$ μmole/cm^4. Assuming this situation and a characteristically more gradual gradient in Mg^{++} of 0.1 μmole cm^{-4}, the

TABLE 3-1

Diffusion coefficients D_{ij} relating flux to concentration gradient for major ions of *seawater* (not sediments). To calculate the flux of a given ion i, use the expression:

$$J_i = -\sum_{j}^{n-1} D_{ij} \frac{\partial C_j}{\partial x},$$

which is summed over all ions (except chloride ion) along a horizontal row in the table. (After Lasaga, 1979.) Units of D_{ij} are 10^{-5} cm^2 sec^{-1}. $T = 25°$C.

Flux	$\frac{\partial C_{SO_4}}{\partial x}$	$\frac{\partial C_{Mg}}{\partial x}$	$\frac{\partial C_{Na}}{\partial x}$	$\frac{\partial C_{Ca}}{\partial x}$	$\frac{\partial C_{K}}{\partial x}$	$\frac{\partial C_{HCO_3}}{\partial x}$	$\frac{\partial C_{PO_4}}{\partial x}$	$\frac{\partial C_{NH_4}}{\partial x}$
J_{SO_4}	1.486	−0.007	−0.013	−0.001	−0.002	0.016	0.018	−0.007
J_{Mg}	0.131	0.890	0.019	0.066	0.003	0.039	0.660	—
J_{Na}	−0.420	0.758	1.546	0.708	0.021	−0.225	−0.800	0.015
J_{Ca}	0.023	0.016	0.004	0.919	0.001	0.017	0.065	—
J_{K}	0.002	−0.032	−0.034	0.021	1.961	—	0.015	—
J_{HCO_3}	0.001	−0.002	—	−0.001	—	1.335	0.002	—
J_{PO_4}	—	—	—	—	—	—	1.634	—
J_{NH_4}	—	—	—	—	—	—	—	1.963

—means value less than .001

data of Table 3-1 enable the calculation:

$$J_{\text{Mg-Mg}} = 8.9 \times 10^{-7}\ \mu\text{mol cm}^{-2}\ \text{sec}^{-}1,$$
$$J_{\text{Mg-SO}_4} = 13.1 \times 10^{-7}\ \mu\text{mol cm}^{-2}\ \text{sec}^{-1},$$

where $J_{\text{Mg-Mg}}$ = flux of magnesium due to the concentration gradient in magnesium;

$J_{\text{Mg-SO}_4}$ = flux of magnesium due to the concentration gradient in sulfate.

Note that under these conditions the flux of magnesium is affected *more* by the concentration gradient in sulfate than it is by the concentration gradient in magnesium itself.

Although electrical effects due to other ions can bring about appreciable changes in the diffusion coefficient of an ion, an even greater effect arises in sediments due to the phenomenon of tortuosity. Tortuosity results from the presence of solid particles. An ion is not free to diffuse in any direction, as it is in the overlying water, but instead is hindered by collisions with the particles as it follows the tortuous path of the fluid between and around them. Mathematically tortuosity is defined as:

$$\theta = d\ell/dx, \tag{3-48}$$

where θ = tortuosity;

$d\ell$ = length of the actual sinuous path over a depth interval dx.

Since $\ell \geqq x$, $\theta \geqq 1$. The diffusion coefficient of a sediment, in terms of tortuosity, is expressed as:

$$D_s = \frac{D}{\theta^2}, \tag{3-49}$$

where D_s = whole sediment diffusion coefficient in terms of area of sediment per unit time.

Since $\theta \geqq 1$, this relation shows that, as expected, sediment diffusivity is lower than that of the pore water alone owing to tortuosity.

It is difficult to calculate tortuosity for real sediments on an *a priori*, geometrical basis. (Some analogous calculations for porous resins are given by Aris, 1975.) Instead, tortuosity is normally determined indirectly. One way to do this is to make measurements of the diffusion coefficient in seawater and in sediments whose pore water composition is virtually the same as seawater. (With this method Li and Gregory (1974) have found the value $\theta^2 = 1.8$ in deep-sea red clay.) Another, easier and more widely used technique is to make measurements of electrical resistivity on natural sediments and on pore waters separated from them. This method has been employed by, e.g., Manheim and Waterman (1974) and Turk (1976). The relation used to calculate tortuosity by this technique is (McDuff and Ellis, 1979):

$$\theta^2 = \phi F, \tag{3-50}$$

where F = formation factor = R/R_0,

and R = electrical resistivity of the sediment,

R_0 = resistivity of pore fluid alone.

McDuff and Ellis (1979) by independently measuring ϕ, R, R_0, D, and D_s have shown that this relation holds very well for the interdiffusion of Br^- and Cl^- in artificial "sediments" consisting of synthetic materials plus KBr-KCl solutions.

An important but simple relation has been found between R/R_0 and porosity for a wide variety of sediments and sedimentary rocks (Manheim, 1970). It is:

$$F = \phi^{-n}. \tag{3-51}$$

If $n = 2$, equation (3-51) is known as Archie's Law, which is a good approximation for sands and sandstones. Results of F and porosity for a variety of sediments collected as part of the Deep Sea Drilling Project are shown in Table 3-2 and Figure 3-11. Note that for sediments of high porosity, values of ϕF (1.4 to 1.8) calculated from electrical resistivity measurements are similar to the results of Li and Gregory for θ^2 in surficial deep-sea red clay (1.8). An average n value for all sites is 1.8 (see Figure 3-11), which agrees well with the independent results of McDuff and Gieskes (1976).

In applying values of D_s to sediments, an additional factor must be included. That is, in order for units of J to come out correctly in Fick's First Law, one needs to multiply D_s by ϕ. In other words, Fick's First Law

TABLE 3-2

Value of the formation factor F for a variety of deep-sea sediments from the Deep Sea Drilling Project (DSDP). (Data from Manheim and Waterman, 1974.) Numbers listed under sediment type represent each DSDP drilling site.

	Depth (*m*)	ϕ	F
Clayey, nanno-plankton ooze (212)	12	0.79	2.0
" "	170	0.48	4.5
" "	299	0.55	4.1
Clayey, diatom ooze (213)	<1	0.87	1.6
" "	15	0.86	1.7
" "	24	0.87	1.9
" "	62	0.84	1.7
Foram-nanno ooze (214)	6	0.68	2.9
" "	63	0.68	2.7
" "	136	0.66	2.6
" "	254	0.47	3.7
" "	330	—	3.7
Radiolarian ooze (215)	6	0.86	1.6
" "	38	0.88	1.7
Nanno ooze (215)	85	0.58	3.0
" "	199	0.55	3.2
" "	115	0.51	3.3
" "	136	0.50	4.2
Silts with interspersed nanno ooze (218)	72	0.53	3.8
" "	232	0.45	5.8
" "	376	0.48	6.8
" "	461	0.41	6.3
" "	488	0.41	7.5
" "	613	0.41	7.9
Clay-rich nanno ooze (217)	7	0.72	2.5

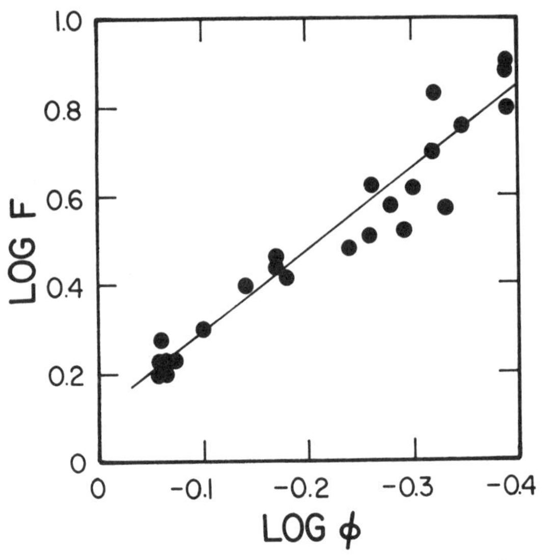

FIGURE 3-11. Plot (log–log) of formation factor F vs porosity for sediments collected by the deep-sea drilling project (Manheim and Waterman, 1974). Data from Table 3-2. Slopes vary from drilling site to drilling site. Data for 5 sites lumped together here, which give an overall slope of $n = 1.8$.

for sediments is:

$$J_s = -\phi D_s \frac{\partial C}{\partial x}, \tag{3-52}$$

where J_s = diffusion flux in sediments in terms of mass per area of *total sediment* per unit time.

Note that if $\hat{C}_d$ (the concentration of a dissolved species per unit volume of total sediment) were substituted for C, there would be no need to introduce ϕ. In this case molecular diffusion as well as bulk diffusive processes such as bioturbation would be represented by the simple diffusive term in the general diagenetic equation (see equation 2-13). However, molecular diffusion occurs only in the interstitial solution and does not respond to a gradient in porosity as do bulk diffusive processes (see Figure 3-16). Use of $\hat{C}_d = \phi C$ for dissolved species would, therefore, be incorrect in that Fick's Law above would include a term involving $\partial\phi/\partial x$, which is physically unreasonable.

On the basis of equation (3-52), Fick's Second Law for sediments is:

$$\frac{\partial C}{\partial t} = -\frac{1}{\phi}\frac{\partial J_s}{\partial x} = \frac{1}{\phi}\frac{\partial\left(\phi D_s \frac{\partial C}{\partial x}\right)}{\partial x}, \tag{3-53}$$

which in expanded form is:

$$\frac{\partial C}{\partial t} = D_s \frac{\partial^2 C}{\partial x^2} + \left(\frac{\partial D_s}{\partial x} + \frac{D_s}{\phi}\frac{\partial \phi}{\partial x}\right)\frac{\partial C}{\partial x}. \tag{3-54}$$

It is often stated that molecular diffusion in sediments is inhibited by rapid adsorption and ion-exchange on the surfaces of the solid particles. This idea, although useful, can lead to much confusion when modeling diagenesis (Manheim, 1970; Berner, 1976). Equilibrium adsorption is a separate chemical process which can be accounted for in diagenetic equations, but whose effects generally do not all appear in the diffusion coefficient. Only when there is (1) no chemical reaction other than equilibrium adsorption and (2) negligible deposition compared to diffusion, can the effects of adsorption all be included in the diffusion coefficient. (See section in Chapter 4 entitled "Equilibrium adsorption and ion exchange" for further details.) This is often the case when waste materials are dumped on sediments (e.g., Duursma and Eisma, 1973), or when sediment diffusion coefficients are measured in the laboratory using Fick's Second Law (e.g., Li and Gregory, 1974; Duursma and Bosch, 1970), but not the normal situation for diagenesis. Furthermore, in Fick's *First* Law it is always incorrect to include the effects of adsorption in the diffusion coefficient.

Some simple reasoning illustrates this latter point. The apparent diffusion coefficient, D', measured in the lab via Fick's Second Law on the assumption of simple linear adsorption, can be expressed as:

$$D' = \frac{D_s}{1 + K}, \tag{3-55}$$

where D' = apparent diffusion coefficient as measured in the laboratory;
K = dimensionless adsorption constant.

Derivation of (3-55) (see Chapter 4 under "Equilibrium adsorption") is based on Fick's First Law where D_s represents the true diffusion coefficient without adsorption. Thus, any use of D' in place of D_s in Fick's First Law is a self-contradiction.

Another factor affecting molecular diffusion, not mentioned so far, is temperature. The higher the temperature the greater are molecular speeds and, thus, the higher is the diffusion coefficient. Seasonal temperature changes can bring about appreciable temperature variations in shallow water sediments (e.g., Aller, 1977), and correction for these changes on the rate of diffusion, if data are available, should be made. Some values of D for ions in seawater and their variation with temperature taken from the work of Li and Gregory (1974) are shown in Table 3-3. The same temperature dependence was found by Li and Gregory for diffusion coefficients measured in sediments.

TABLE 3-3

Tracer diffusion coefficients for some ions in seawater. (Tracer diffusion refers to self-diffusion of radioactive isotopes, which minimizes the electrical effects of other ions.) (Data from Li and Gregory, 1974.)

		(10^{-6} cm^2 sec^{-1})			
Temperature °C	Na^+	Ca^{++}	K^+	Cl^-	SO_4^{--}
23.7 ± 0.4	13.4	7.5	17.9	18.6	9.8
5.0 ± 0.3	8.0	5.0	11.4	11.5	5.8

From the data of Tables 3-1 and 3-3 and resistivity measurements or formation factor estimates (for example from porosity determinations), one may calculate the diffusive flux or the diffusion coefficient of Na^+, Ca^{++}, K^+, Cl^-, and SO_4^{--} in a given sediment. However, as a service to the reader for the purpose of quick estimates, actual measured values are listed in Table 3-4 for some diagenetically important ions in typical fine-grained muds. Note the relative insensitivity of the value of D_s for sulfate upon electrical interaction, as predicted by Table 3-1. (Tracer diffusion refers to the self-diffusion of a radioactive tracer, and involves minimal electrical coupling due to a lack of concentration gradients of other ions.)

TABLE 3-4

Diffusion coefficients for some typical fine-grained marine sediments (T = 20–25°C). Coupling refers to other ions which have large concentration gradients and thus maximal electrical interactions with the ion of interest.

Ion	ϕ	D_s (10^{-6} cm^2 sec^{-1})	*Coupling*	*Reference*
Na^+	0.71	7.4	none (tracer)	(1)
Ca^{++}	0.71	4.4	" "	(1)
Cl^-	0.71	10.2	" "	(1)
SO_4^{--}	0.71	5.0	" "	(1)
SO_4^{--}	0.72	5.0	with HCO_3^-	(2)
SO_4^{--}	0.64	4.0	with HCO_3^-	(3)
NH_4^+	0.72	9.8	with HCO_3^-, SO_4^{--}	(2)
HPO_4^{--}	0.72	3.6	with HCO_3^-, SO_4^{--}	(2)

(1) Li and Gregory (1974).
(2) Krom and Berner (1980).
(3) Goldhaber et al. (1977).

Benthic Boundary Diffusion

Near the sediment-water interface, sediments are affected by physical processes occurring in the overlying water. The most obvious process is resuspension of particles by sudden increases in near-bottom flow velocity, which is often referred to as wave and current stirring. Although sediment resuspension and transport above the bottom are important processes, they are complex, and detailed discussion of them is beyond the scope of this book. The interested reader is referred to summary works on the subject such as that of Allen (1970). Nevertheless, one can describe the average homogenizing effects of resuspension on sediment properties near the sediment-water interface (in the absence of more detailed knowledge) in terms of an eddy diffusion coefficient. This has been done by Vanderborght et al. (1977a) to describe the concentration of dissolved silica in sediments from the North Sea. Here they assumed a constant eddy diffusion coefficient 100 times larger than that for molecular diffusion down to 3.5 cm depth (see Chapter 6 for further details). The formal treatment of wave and current stirring according to this approach is in terms of Fick's Second Law for simultaneous mixing of both solids and pore water. This is (for a constant diffusion coefficient):

$$\frac{\partial \hat{C}}{\partial t} = D_{wc} \frac{\partial^2 \hat{C}}{\partial x^2}, \tag{3-56}$$

where $\hat{C}$ = concentration of a dissolved or solid species per unit volume of total sediment;

D_{wc} = wave and current diffusion coefficient in area of sediment per unit time.

For a dissolved constituent with concentration, C:

$$\frac{\partial(\phi C)}{\partial t} = D_{wc} \frac{\partial^2(\phi C)}{\partial x^2}. \tag{3-57}$$

Note that here, in contrast to molecular diffusion, homogenization can occur in response to porosity gradients as well as to concentration gradients. Equation (3-57), with constant porosity, is the actual expression used by Vanderborght et al. (1977) to describe the concentration of dissolved silica.

Even when bottom water flow is relatively constant and there is no sediment resuspension (or bioturbation), there still may be an effect of the bottom water on dissolved constituents in a sediment. This has to do with the presence of the diffusive sublayer of the benthic boundary layer (Morse, 1974; Boudreau and Guinasso, 1980). Ions migrating between the sediment and the overlying water must pass through a thin layer of water adjacent to the sediment where slow diffusion occurs via molecular processes as compared to the overlying water where rapid homogenization is attained via turbulent mixing. This diffusive sublayer results from the fact that velocity fluctuations (which give rise to turbulent diffusion) decrease and approach zero as the sediment surface is approached. As a result, molecular processes become quantitatively more important than turbulence in bringing about the transfer of dissolved constituents. If rates of diagenetic reaction within the sediment are high, concentrations of dissolved species near the sediment-water interface may build up and bring about a rise in concentration within the diffusive sublayer above the value found for the overlying (well-mixed) water. In this case the concentration at $x = 0$ would not be the value for the overlying water, and this would have to be taken into consideration when constructing diagenetic models (Morse, 1974).

One of the major problems with the diffusive sublayer is the matter of its thickness. At typical deep-sea bottom current velocities, Boudreau and Guinasso (1980) have calculated, assuming a perfectly flat sediment-water interface, a thickness of about one millimeter. However, the sediment-water interface is rarely flat, due to irregularities brought about by irregular sedimentation or bioturbation (e.g., burrow mounds), and the question arises as to whether a one-millimeter thickness is correct. The problem is further compounded by observation of sporadic bursting (sudden release of ions on sediment particles) through thin layers (Heathershaw, 1974) which, if frequent enough, would require modification of the simple picture of molecular diffusion. Finally, bioturbation may be more rapid than either diffusive sublayer transfer or wave and current stirring, especially in near-shore sediments. In this case the concept of a diffusive sublayer loses its significance.

The major process of benthic boundary diffusion is bioturbation. Because of its uniqueness and importance it is discussed as a separate topic in the next section.

Bioturbation

Sediments which have not been buried appreciably, say less than about one meter, are subject to mixing by the activities of benthic organisms. The resulting process is known as bioturbation. (For a detailed discussion of this subject consult Rhoads 1974 or Aller, 1977; 1980. A generalized diagram taken from the latter references is shown in Figure 3-12.) Bioturbation occurs in several different ways. Some organisms, such as crabs and snails, mix surface sediment simply by crawling or plowing through it.* More importantly, others, especially polychaete worms and bivalves,

* A subtype of this kind of bioturbation is that brought about by sedimentologists while sampling bottom sediments, i.e., anthropoturbation.

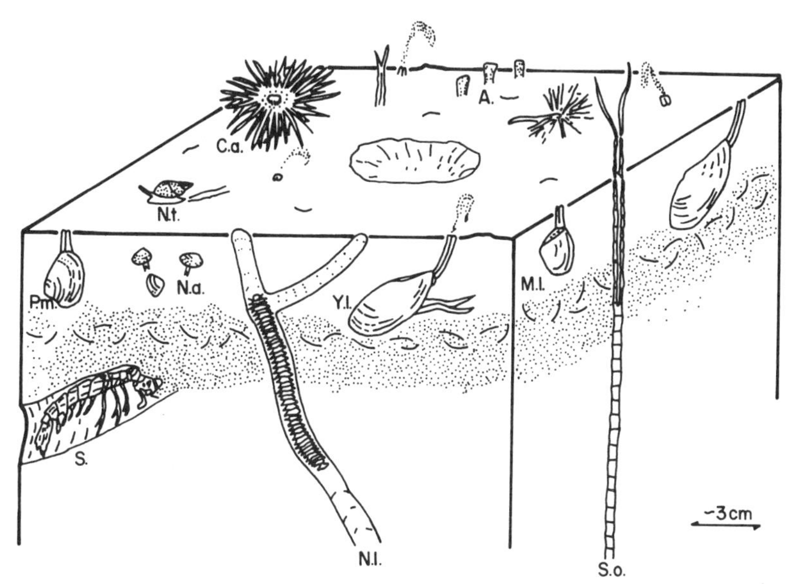

FIGURE 3-12. Schematic representation of major burrowing fauna from a sediment locality (NWC) in Long Island Sound. (From Aller, 1977; 1980.)

burrow into sediment and ingest the sediment particles. Such burrowing can extend to several tens of centimeters. Once their burrows are constructed, some organisms remain in them and flush the burrows with overlying seawater, thus bringing about enhanced exchange between pore waters and overlying seawater. This process is referred to as irrigation; it involves only the pore water and not the enclosing particles.

Exact mathematical modeling of bioturbation is exceedingly difficult because of the variety, irregularity, and complexity of the various bioturbational processes. The usual approach has been to lump all processes together and describe bioturbation simply as a random mixing process. (Goldberg and Koide, 1962; Berger and Heath, 1968; Hanor and Marshall, 1971; Guinasso and Schink, 1975; Peng et al., 1977; Nozaki, et al., 1977). This is done in two basic ways. The simplest approach (e.g., Berger and Heath, 1968) is to assume such fast mixing over a certain depth that all sediment properties are uniform from the sediment-water interface down to this fixed depth of mixing. In other words, a "box-model" approach is adopted (see Figure 3-13). According to this model, changes in the properties of sediment added at the sediment-water interface are immediately sensed at the bottom of the "box" but are damped by the mixing. For situations involving solids where there are no diagenetic changes other than bioturbation, the proper mathematical representation of this

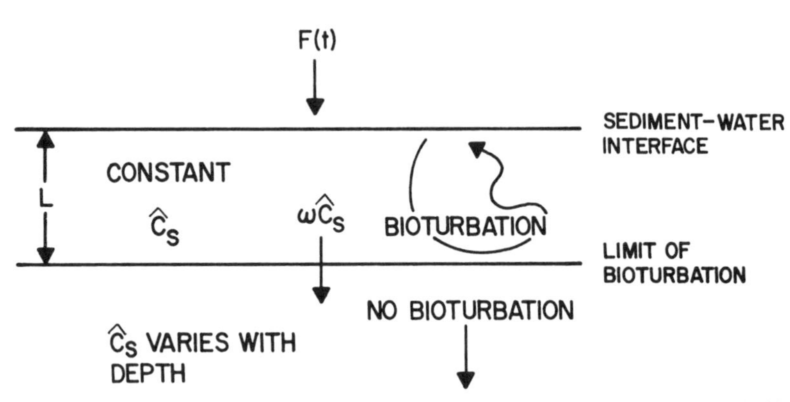

FIGURE 3-13. Schematic representation of box model normally used to quantify bioturbation. Within the box there is perfect mixing and uniformity of concentration.

type of box model (Guinasso and Schink, 1975) is:

$$\frac{d\hat{C}}{dt} = \frac{F(t) - \omega\hat{C}_s}{L}, \tag{3-58}$$

where $\hat{C}_s$ = concentration in the zone of bioturbation (box) of solid substance under study in terms of mass per unit volume of total sediment (solids plus pore water);

t = time;

$F(t)$ = depositional flux to sediment surface of substance under study (mass area^{-1} time^{-1});

ω = rate of depositional burial of total sediment (compactive flow of pore water is ignored);

L = thickness (depth) of box which is assumed to be constant with time.

Because there is no other diagenetic alteration, the record of depositional changes, as modified by bioturbation, is fixed at the time the sediment passes downward by burial below the zone of bioturbation. Thus, knowledge of $F(t)$ can be used via solution of (3-58) to predict the buried record, or, more importantly, the buried record can be used to deduce $F(t)$.

Through the use of equation (3-58) Berger and Heath (1968) and Ruddiman and Glover (1972) show how box modeling can be used to describe the record left after bioturbational mixing of an impulse source.

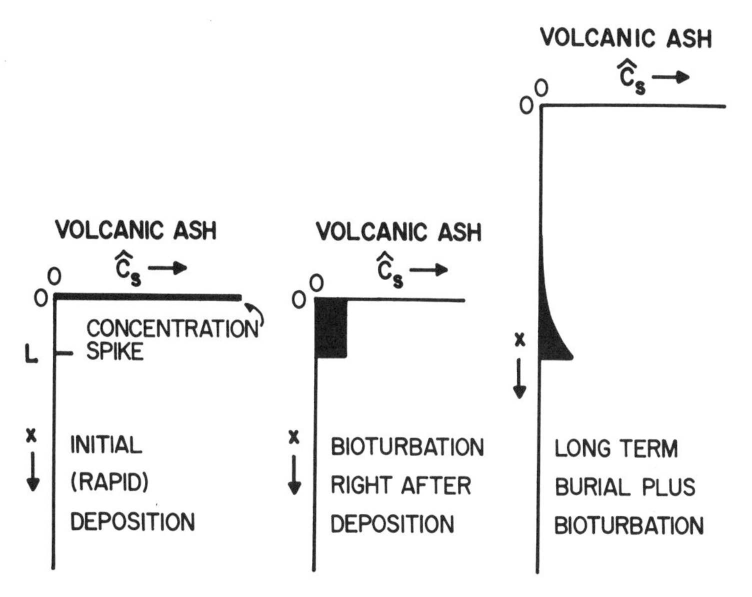

FIGURE 3-14. Concentration distribution (schematic) of volcanic ash particles in sediments undergoing box-model-type bioturbation. Area under concentration curve darkened. The record of the original "event" is preserved as an upward exponentially decreasing curve of ash particle concentration. (Modified from Ruddiman and Glover, 1972.)

A good example of an impulse source is the sudden deposition of volcanic material from an ashfall (Ruddiman and Glover, 1972). Without bioturbation the ashfall would be represented by a thin layer with sharp upper and lower boundaries. With box-model-type bioturbation the ash particles, upon deposition, are immediately mixed downward into the underlying sediment, and then subsequently mixed into overlying sediment as it is deposited. As a result the lower contact remains sharp, but reduced in contrast, while the upper boundary becomes gradational. This is illustrated in Figure 3-14.

Box modeling can also be used to describe bioturbational mixing of materials undergoing chemical reaction or radioactive decay. An excellent example is provided by the work of Nozaki et al. (1977) and Peng et al. (1977) for the distribution of ^{14}C in pelagic sediments. The results of the former study are shown in Figure 3-15. Because of bioturbation and mixing with older material the apparent ^{14}C age of $CaCO_3$ particles at the surface of a pelagic sediment is much greater than the modern age expected, and the difference can be predicted using steady-state box modeling. This is shown by the following calculation which is patterned after those of Nozaki et al. and Peng et al.

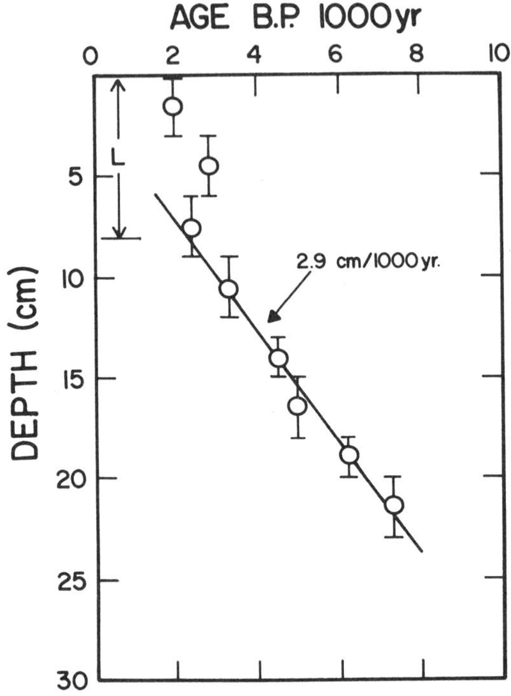

FIGURE 3-15. Age via ^{14}C dating of sediments vs depth in a core from the Mid-Atlantic Note old surface age due to bioturbation. The depth of bioturbation (deduced from the break in slope) is denoted by L (8 cm). (After Nozaki et al., 1977.)

The input of $CaCO_3$-bound ^{14}C to the mixed layer or "box" per unit area is due to sedimentation alone:

$$\text{Input} = (^{14}C/^{12}C)_{\text{input}}(N\bar{\rho}_s/MW)(1-\phi)\omega, \tag{3-59}$$

where $(^{14}C/^{12}C)$ = atomic ratio of carbon isotopes in $CaCO_3$;
N = mass fraction of $CaCO_3$ to total solids;
MW = molecular weight of $CaCO_3$;
$\bar{\rho}_s$ = average density of total solids;
ϕ = porosity;
ω = sedimentation rate (compaction is ignored).

The output is due to burial and radioactive decay (assuming no dissolution of $CaCO_3$):

$$\text{Output} = (^{14}C/^{12}C)_{ML}(N\bar{\rho}_s/MW)(1-\phi)\omega + \lambda(^{14}C/^{12}C)_{ML}(N\bar{\rho}_s/MW)(1-\phi)L, \tag{3-60}$$

where λ is the decay constant of ^{14}C and L is the thickness of the mixed layer (ML = mixed layer). At steady state input is equal to output, or upon equating expressions (3-59) and (3-60) and solving for $(^{14}C/^{12}C)_{ML}$,

$$(^{14}C/^{12}C)_{ML} = (^{14}C/^{12}C)_{\text{input}} \left(\frac{\omega}{\omega + \lambda L} \right). \tag{3-61}$$

Now by radioactive decay:

$$(^{14}C/^{12}C)_{ML} = (^{14}C/^{12}C)_{\text{input}} \exp(-\lambda T), \tag{3-62}$$

where $T = {}^{14}C$ "age" of the mixed layer. Substituting (3-62) in (3-61) and solving for T:

$$T = \frac{1}{\lambda} \ln \left(1 + \frac{\lambda L}{\omega} \right). \tag{3-63}$$

In Figure 3-15 the value of ω (2.9 cm/10^3 yr) can be readily deduced from the slope of age vs depth below the mixed zone of constant age. Using this and the thickness of the mixed zone ($L = 8$ cm), Nozaki et al. (1978) calculated the ^{14}C "age" of the mixed zone, using equation (3-63), to be 2,400 years, which is in excellent agreement with that actually measured. Similarly good agreement was found by Peng et al. for other pelagic sediments. Thus, simple steady state box-modeling appears to describe successfully the bioturbation of $CaCO_3$-bound ^{14}C in deep-sea sediments.

Box models necessarily assume that mixing is so rapid that it brings about constant homogenization within the zone of bioturbation. However, the mixing rate is not truly infinite, and the choice of whether or not to employ a box model depends upon the time scale of the process under study. Apparently the radioactive decay of ^{14}C is sufficiently slow (half life of 5,730 years) that benthic mixing in the deep sea can maintain homogeneity. By contrast, more rapid processes do not lend themselves to a box-model description. For instance Nozaki et al. found that the radionuclide ^{210}Pb (which is restricted to particles because of the low solubility of lead), did not show uniformity in concentration over the box-model mixing depth. Also, the gradient found for ^{210}Pb did not reflect simple burial since, at the known rate of sedimentation, 0.003 cm per year, all ^{210}Pb should be restricted to the top millimeter. (Its half life is only 22 years.) Thus, the ^{210}Pb distribution is describable neither by box-modeling nor by simple layer-by-layer burial. A different type of model is necessary.

Bioturbation which does not result in complete homogenization, but which can still be treated as random mixing, is best described in terms of a biological mixing or "biodiffusion" coefficient. This approach has been adopted in the models of Goldberg and Koide (1962), Guinasso and Schink (1975), Nozaki et al. (1977; for ^{210}Pb), and Aller (1977; 1980). (The use of biotransport by Hakanson and Kallstrom (1978) can be shown to be equivalent to this approach except that the biodiffusion coefficient, in this case, changes with depth.) In applying biodiffusion coefficients one should be careful to distinguish particle bioturbation from fluid bioturbation, because the two often take place at distinctly different rates because of irrigation. This is emphasized and actually demonstrated by Aller (1977) for near-shore sediments.

Mathematical treatment of *solid* biodiffusion requires the introduction of the biodiffusional flux:

$$J_B = -D_B \frac{\partial \hat{C}_s}{\partial x}, \tag{3-64}$$

where $\hat{C}_s$ = mass of biodiffusing solid under study per unit volume of total sediment (solids plus water);

J_B = biodiffusional flux of solids in mass per unit area of sediment per unit time;

D_B = biodiffusion coefficient for solids in terms of area of total sediment squared per unit time.

For pore water, in order to account for the separate process of irrigation, we introduce an additional parameter, the irrigation (biodiffusion) coefficient so that:

$$J_{BI} = -D_B \frac{\partial(\phi C)}{\partial x} - \phi D_I \frac{\partial C}{\partial x}, \tag{3-65}$$

where C = mass of biodiffusing dissolved constituents per unit volume of pore water;

J_{BI} = biodiffusional flux of dissolved solute;

D_I = irrigation coefficient in terms of area of total sediment squared per unit time;

ϕ = porosity.

Note that D_B is included in the expression for pore water biodiffusion because solid biodiffusion, if it is a true random mixing process, unavoidably involves accompanying pore water biodiffusion. To avoid semantic confusion we will henceforth refer to bioturbation in terms of particle mixing, D_B (which includes pore water), and irrigation, D_I, of pore water.

Insertion of these flux expressions into the general diagenetic equation poses no problem. They are simply combined with the flux from molecular diffusion to give (ignoring advection and chemical reaction):

$$\frac{\partial \hat{C}_s}{\partial t_{diff}} = \frac{\partial \left(D_B \frac{\partial \hat{C}_s}{\partial x} \right)}{\partial x}, \tag{3-66}$$

$$\frac{\partial(\phi C)}{\partial t_{diff}} = \frac{\partial \left[D_B \frac{\partial(\phi C)}{\partial x} + \phi(D_s + D_I) \frac{\partial C}{\partial x} \right]}{\partial x}. \tag{3.67}$$

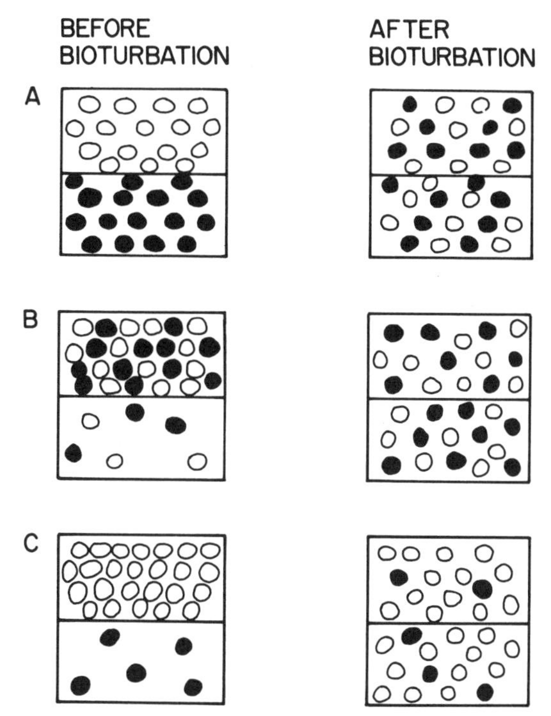

FIGURE 3-16. Schematic representation of bioturbational mixing of both solids and pore water. The dark and light "balls" represent two different types of solid particles. The space between balls represents interstitial water.

(A) Homogeneous distribution of porosity; inhomogeneous distribution of particle types relative to one another.

(B) Inhomogeneous distribution of porosity; homogeneous distribution of particle types relative to one another

(C) Inhomogeneous distribution of both porosity and particle types.

Note in all three cases that $\partial \hat{C}_s / \partial x \neq 0$ for black (or white) balls, and that bioturbation brings about homogenization of both balls and water.

Note that biodiffusion of a dissolved constituent accompanying particle mixing is treated differently from molecular diffusion or irrigation, in that flux of a dissolved constituent during particle mixing *can* result solely from a gradient in porosity, whereas molecular diffusion or irrigation fluxes cannot. In other words, porosity (water content) is a parameter subject to particle biodiffusion as well as concentrations of dissolved and solid constituents. This is all shown in Figure 3-16.

In any given situation the relative magnitude of each of the three diffusion coefficients in equation (3-67) determines whether any one (or two) of them can be ignored. In deep-sea sediments, it has been abundantly

TABLE 3-5

Values of D_B, the biodiffusion coefficient for solid particles for various marine sediments.

Sediment Type	*ω in cm/1000 yr*	D_B 10^{-9} cm^2 sec^{-1}	*Method and Literature Source*
Near-shore slit clay	100–300	200	^{234}Th, (1) (2)
" " "	"	1200	" " "
" " "	"	150	" " "
" " "	—	500	" " "
Pelagic calcareous ooze (Atlantic Ocean)	2.9	6	^{210}Pb (2)
" " "	—	2	" "
Clay-siliceous ooze (Equat. Pacific)	0.15	14	" "
" "	0.31	8	" "
" "	0.14	10	" "
Pelagic siliceous ooze (Antarctic)	1	1	" "
" " "	1	7	" "
" " "	1	8	" "
" " "	4	2	" "
Deep-sea, pelagic sediments	0.51	0.04	Tektites (3)
" "	0.75	0.02	" "
" "	0.74	>4	" "
" "	0.46	0.02	" "
" "	0.22	0.02	" "
" "	1.51	0.1	" "
" "	~1.0	>6	" "
" "	0.64	0.02	" "
Deep-sea, pelagic sediments	—	12	Plutonium (3)
" "	—	7	" "
" "	—	8	" "
" "	—	4	" "
" "	—	3	" "

(1) Aller (1977).
(2) Turekian et al. (1978).
(3) Guinasso and Schink (1975).

shown that D_B is much less than D_s (see Table 3-5). However, the magnitude of D_I in such sediments has not been determined, and it needs to be demonstrated that D_I is in fact negligible compared to D_s as is usually assumed. In near-shore, organic-rich sediments Aller (1977; also see below) has shown that D_I is much greater than D_s or D_B. (In units of cm^2 sec^{-1} $D_I \approx 15 \times 10^{-6}$, $D_s \approx 3 \times 10^{-6}$, and $D_B = 0.1\text{–}1.0 \times 10^{-6}$).

For the sake of mathematical simplicity it is usually assumed that D_B (and D_I) is constant down to a fixed depth of bioturbation and that below this depth pore water diffusion takes places entirely by molecular processes and solids are not mixed. This gives rise to two diagenetic equations, one for the zone of bioturbation (with constant D_B and D_I) and another for the sediments below it (constant D_s), with continuity of concentration and flux maintained at the boundary between the two regimes. Modeling of this sort, when applied to solid particles in deep-sea sediments has been, in general, rather successful. For example, Guinasso and Schink (1975) have modeled tektite biodiffusion from buried, originally thin layers of known age in deep-sea sediments and calculated values of D_B which agree well with one another considering the crudeness of the assumptions (see Table 3-5). In addition, biodiffusional modeling of radio-tracers present in the upper few centimeters of deep-sea sediments provide values of D_B which are in relatively good agreement. A summary of D_B data for various sediments is given in Table 3-5. Note that the distinctly lower values calculated for buried tektite layers can be explained by longer term exposure to rare, deep mixing events.

In near-shore sediments, particle bioturbation is much more intense than in the deep sea due to higher concentrations of organic material (food) and higher temperatures. This results in higher values for D_B. Also, seasonal temperature variation results in seasonal variation in D_B. This is shown by the results of Aller (1977) for sediments from Long Island Sound, U.S.A (Table 3-5).

Aller (1977; 1980) has also studied bioturbation of pore water in near-shore sediments in great detail and has found it to exhibit seasonal variations and to be much more rapid than particle bioturbation. He has suggested that enhanced pore water bioturbation is best described in terms of the irrigation of burrows, and on this basis has proposed a new irrigation model to describe this process. His model (which is discussed in more detail in Chapter 6), briefly summarized, proceeds as follows: At any given depth vertical cylindrical burrows are uniformly spaced and flushed so rapidly that water in them has the same composition as seawater overlying the sediments. This is in keeping with observations of natural burrows. Between burrows, at any given depth, dissolved species produced by diagenetic reactions migrate horizontally by molecular diffusion to the nearest burrow where the concentration is held at the low value of the overlying seawater. Assuming cylindrical symmetry (see equation 3-42), the appropriate diagenetic equation expressing both cylindrically symmetric and vertical diffusion is:

$$\frac{\partial C}{\partial t} = \frac{D_s}{r}\frac{\partial\left(r\frac{\partial C}{\partial r}\right)}{\partial r} + D_s\frac{\partial^2 C}{\partial x^2} - \omega\frac{\partial C}{\partial x} + R, \tag{3-68}$$

where r = horizontal radial coordinate with center of burrow as origin;

R = rate of reaction which is assumed to be a function of depth x, but not of radial distance;

D_s = molecular diffusion coefficient.

Assuming steady state ($\partial C/\partial t = 0$) and ignoring burial advection, Aller integrated equation (3-68) and calculated the average concentration for each sampling depth interval in the sediment from the expression:

$$\bar{C} = \frac{2\pi\int_{x_1}^{x_2}\int_{r_1}^{r_2} Cr\,dr\,dx}{2\pi\int_{x_1}^{x_2}\int_{r_1}^{r_2} r\,dr\,dx}, \tag{3-69}$$

where $\bar{C}$ = average concentration in pore water at each depth;

r_1 = radius of burrow;

r_2 = one-half the average distance between burrows;

x_1, x_2 = lower and upper boundaries of sampling depth interval.

Both r_1 and r_2, as stated above, are functions of depth and were obtained from actual measurements.

Using measured values of D_s and R, Aller calculated a profile of $\bar{C}$ vs x for steady state which could be well fitted to measured concentrations. He then applied a simple one-dimensional biodiffusion model (like that discussed above) to the same problem. In other words,

$$\frac{\partial \bar{C}}{\partial t} = D_I\frac{\partial^2 \bar{C}}{\partial x^2} + R. \tag{3-70}$$

This equation was then solved for steady state and $\bar{C}$ vs depth profiles calculated for various values of D_I. In order to obtain average concentrations at each depth comparable to those calculated according to the radial model, Aller had to raise D_I to a value about 4 times higher than that used for D_s in the radial model. Concentrations calculated via the one-dimensional model for $D_I = D_s$ were far too high. From this he concluded that (1977, p. 336) "radial diffusion into burrows is capable of significantly lowering pore water concentrations below that predicted by only vertical diffusion alone." In addition, he found that the shape of the profiles calculated according to the radial model more closely approximated those actually measured. (Average concentrations were obtained by lateral homogenization of sediment at each depth in box cores.) Overall it appears that Aller's burrow model is a more realistic representation of irrigation, in that resort to highly idealized parameters, such as D_I, is unnecessary, and the process can instead be described in terms of molecular diffusion across a highly invaginated (due to burrowing) sediment-water interface.

Some other new models, which do not rest upon the assumption of random mixing to describe bioturbation, are those of Hakanson and Kallstrom (1978) and Goreau (1977). Hakanson and Kallstrom assume that bioturbation occurs via vertical advection (biotransport) of total sediment, and model the porosity distribution in the top 10–20 cm of lake sediments in terms of compaction brought about by divergence of the biotransport flux. Such an advective approach is necessary when describing bioturbation by some organisms which, due to selective feeding and defecation of fine grains, actually bring about unmixing and segregation of grain types into distinct layers (Rhoads, 1974). Goreau (1977) uses signal theory to describe how bioturbation converts an original concentration fluctuation at the sediment-water interface to final concentration-vs-depth profiles. These studies, as well as that of Aller, show that different theoretical approaches can be used in the study of bioturbation, and that treatment in terms of biodiffusion is merely a first step.

Transfer across the Sediment-Water Interface

One of the most important consequences of early diagenesis is the control it exerts upon the chemical composition of natural waters. If diagenetic chemical reactions occur close enough to the sediment-water interface, sharp concentration gradients result, and because of diffusion and advection, fluxes of dissolved species between sediment and overlying water occur. These fluxes can have a major effect on the composition of the overlying water. For example, Sayles (1979) has shown that diagenetic reactions in deep-sea sediments followed by molecular diffusion, adds or subtracts Mg^{++}, K^{+}, Ca^{++}, and HCO_3^{-} from seawater at rates that are of the same order of magnitude as the rates by which these species are added to the oceans by rivers. Thus it is important to be able to calculate rates of transfer across the sediment-water interface.

The usual approach to calculating fluxes of dissolved species at the sediment-water interface is to treat all mixing processes in terms of a diffusion coefficient (e.g., Berner, 1980; Aller, 1980; Sayles, 1979; Lerman, 1979) which is used in conjunction with Fick's First Law of diffusion for sediments (equation 3-52). To this is added the advective burial of pore water. Ignoring the effects of porosity gradients on biodiffusion and wave and current stirring, the appropriate expression is:

$$J_0^* = -\phi_0 D_{T_0}\left(\frac{\partial C}{\partial x}\right)_0 + \phi_0 v_0 C_0, \tag{3-70a}$$

where J^* = total (advective plus diffusive) flux;

$D_T = D_s + D_B + D_I + D_{wc}$;

and the subscript zero refers to the sediment-water interface. If steady state compaction is present and sedimentation rate is expressed in terms of mass flux $\mathscr{R}$, one obtains from equations (3-20), (3-24), and (3-70a):

$$J_0^* = -\phi_0 D_{T_0}\left(\frac{\partial C}{\partial x}\right)_0 + \frac{\mathscr{R}}{\bar{\rho}_s}\left(\frac{\phi_x}{1-\phi_x}\right)C_0. \tag{3-70b}$$

In general the advection term is much smaller than the diffusion term and is, thereby, usually omitted (e.g., see Sayles, 1979).

In most studies bioturbation and wave and current stirring are ignored and D_T set equal to D_s, the molecular diffusion coefficient. This is most justifiable when dealing with deep-sea sediments where disturbance of the sediment-water interface is at a minimum. By contrast, Vanderborght et al. (1977a) found that in the upper 3.5 cm of shallow water sediments of the North Sea, D_T, as a result of wave and current stirring, was about 100 times higher than that expected for molecular diffusion. Obviously in situations like this D_T cannot be equated with D_s since the flux across the sediment-water interface depends upon transport properties in the uppermost portions of the sediment. Furthermore, the work of Aller (1977; 1980) shows that enhanced transport of dissolved species must be present where burrow irrigation is widespread. (However, Aller also showed that if most diagenetic reaction occurs in the uppermost centimeter, flux to the overlying water is accomplished mainly by vertical transport and not by radial diffusion into burrows.) Thus, when using equation (3-70b) one must be careful to estimate D_{T_0} properly.

Another problem in calculating fluxes is accurate determination of $(\partial C/\partial x)_0$. Its computed value depends upon the scale of sampling. Sharp gradients in the upper few centimeters can be missed if depth sampling is insufficiently closely spaced. Also, the value of C at the sediment-water interface (C_0) may not be equal to that in the overlying water if a benthic boundary diffusive sub-layer exists (Morse, 1974; Boudreau and Guinasso,

1980). A further problem is that in the presence of appreciable irrigation, with resultant laterally inhomogeneous concentrations, it becomes difficult to specify $(\partial C/\partial x)_0$. In this case it may be preferable to calculate J_0^* using equations different from (3-70b). According to the irrigation model of Aller, the expression for J_0^* would be much more complex and would involve integration of different radial fluxes to burrows over the depth of burrowing.

Summary of Mathematics of Diagenetic Physical and Biological Processes

From what has been discussed in this chapter, one can summarize the effects of physical and biological processes on early diagenesis in terms of expanded forms of the general diagenetic equation. To do this we express the concentration of a solid component in terms of C_s (mass per unit mass of total solids) such that:

$$\hat{C}_s = (1 - \phi)\bar{\rho}_s C_s. \tag{3-71}$$

For a dissolved component we express the concentration in terms of mass per unit volume of pore water such that:

$$\hat{C}_d = \phi C. \tag{3-72}$$

Substituting into the general diagenetic equation (2-13) we obtain:

For a solid component

$$\frac{\partial[(1 - \phi)\bar{\rho}_s C_s]}{\partial t} = \frac{\partial\left\{D_B \frac{\partial[(1 - \phi)\bar{\rho}_s C_s]}{\partial x}\right\}}{\partial x} - \frac{\partial[(1 - \phi)\omega\bar{\rho}_s C_s]}{\partial x} + (1 - \phi)\bar{\rho}_s \sum R_s, \tag{3-73}$$

For a dissolved component

$$\frac{\partial(\phi C)}{\partial t} = \frac{\partial\left[D_B \frac{\partial(\phi C)}{\partial x} + \phi(D_I + D_s)\frac{\partial C}{\partial x}\right]}{\partial x} - \frac{\partial(\phi v C)}{\partial x} + \phi \sum R_d, \tag{3-74}$$

where $\sum R_s$ = all diagenetic reactions affecting C_s;
$\sum R_d$ = all diagenetic reactions affecting C.

*See (4-52) for eq. + adsorption factor

Here chemical reactions which appreciably affect porosity are ignored. Also, the wave and current mixing coefficient D_{wc} is incorporated into the biodiffusion mixing coefficient D_B because of the common difficulty in distinguishing between the two processes in the upper few centimeters of sediment. (For most sediments $\bar{\rho}_s \neq f(x,t)$ since most common sedimentary minerals have about the same density. In this case the parameter $\bar{\rho}_s$ can be dropped from equation 3-73.) These two equations form the basis for most further discussion of early diagenesis in the present book.

4

Diagenetic Chemical Processes I: Equilibrium, Homogeneous, and Microbial Reactions

In the next two chapters we will consider the various chemical processes that go together to make up the ΣR term of the general diagenetic equation. These processes are important in that they constitute the driving force for most diagenetic change in sediments. Subjects discussed are equilibrium processes (including rapid adsorption and ion exchange), homogeneous reactions (including radioactive decay), microbial (metabolic) reactions, precipitation and dissolution, and authigenic processes. The purpose of these two chapters is not to discuss specific reactions themselves, but rather to present general principles from which mathematical representation of each reaction type can be made. As a result, heavy emphasis will be placed on the fields of chemical kinetics and thermodynamics. In order to break up the discussion into more tractable units, we will treat only equilibrium processes and homogeneous and microbial reactions in this chapter. In the following chapter, emphasis will be on mineral reactions, specifically precipitation, dissolution, and authigenic processes.

Equilibrium Processes

If diagenetic chemical reactions are so rapid that chemical equilibrium is essentially maintained in the face of advection and diffusion, then a thermodynamic approach to diagenetic chemical reactions is possible. It is assumed that at each point in space equilibrium is maintained and that all chemical reactions occur at equilibrium. This is the approach used, for example, in the mass transfer weathering models of e.g., Helgeson et al. (1969), Fouillac et al. (1977), and Fritz (1975). Of course, strictly speaking, no net reaction can occur *at equilibrium*, but the assumption is that there is so little kinetic impediment to the reactions that they occur rapidly at very small departures from equilibrium, for example, in the case of dissolution at unmeasurably low degrees of undersaturation. In this sense it is therefore possible to speak of "equilibrium reactions."

Determination of Equilibrium

In order to employ the concept of chemical equilibrium in diagenetic models, one needs to know how to determine equilibrium concentrations.

For the chemical reaction:

$$\alpha A + \beta B \leftrightharpoons \gamma C + \delta D,$$

one can write:

$$K = \frac{a_C^{\gamma} a_D^{\delta}}{a_A^{\alpha} a_B^{\beta}}, \tag{4-1}$$

where K = thermodynamic equilibrium constant, a function of temperature and total pressure only;

a = activity of each product and reactant;

and, the double arrows of the reaction represent reversible equilibrium.

One can calculate the value of K from tabulations of standard free energy data for geological substances (Garrels and Christ, 1965; Robie et al., 1978; Helgeson et al., 1978) according to the relation:

$$\Delta G^{\circ} = -RT \ln K, \tag{4-2}$$

where ΔG° = standard state (unit activity) Gibbs free energy change for the reaction;

R = gas constant;

T = absolute temperature.

Because we deal in this book only with shallow burial where temperatures generally do not exceed 35°C, one often can directly use standard free energy data, which are normally tabulated for 25°C and 1 atm, to calculate K via equation (4-2) if only a rough result is desired. For more accurate results, corrections for temperature and pressure differences must be made. For temperature correction, we use standard enthalpy and heat capacity data (see references above) and the expression:

$$\left(\frac{\partial \ln K}{\partial T}\right)_P = \frac{\Delta h^{\circ}}{RT^2}, \tag{4-3}$$

where Δh° = standard state partial molar enthalpy change for the reaction. For pressure correction (pressures for buried deep-sea sediments may approach 1,000 bars), we use partial molar volume and compressibility data (Millero, 1979, and the above references) and the expression:

$$\left(\frac{\partial \ln K}{\partial P}\right)_T = \frac{-\Delta v^{\circ}}{RT}, \tag{4-4}$$

where Δv° = partial molar volume change for the reaction at the standard state (25°C, pressure of the reaction, and unit activity);

P = pressure.

Once the equilibrium constant K is determined, we still need to convert activities of products and reactants to concentrations, since concentration is what is used in diagenetic equations. This is done through the use of activity coefficients. Activity of dissolved species in interstitial solution is given by the conventional expression:

$$a = \gamma m, \tag{4-5}$$

where γ = conventional (or practical) molal activity coefficient;
m = molality (moles per kg of H_2O).

The value of γ approaches one as the solution becomes more and more dilute. In terms of concentration per unit volume of pore solution C, which is the normal unit used in diagenetic equations:

$$a = (\gamma/\rho_w^*)C, \tag{4-6}$$

where ρ_w^* = mass of water (H_2O) per unit volume of interstitial solution.

For solid solutions (e.g., minerals), and H_2O in pore solution, we adopt the convention:

$$a = \lambda X, \tag{4-7}$$

where λ = rational activity coefficient;
X = mole fraction of a given end-member in a solid solution or water in interstitial solution.

For the major components of a solid solution, the property of λ is such that as X approaches 1, λ approaches 1. For species adsorbed on the surfaces of solids, we here adopt the special convention:

$$a = \psi\bar{C}, \tag{4-8}$$

where $\bar{C}$ = mass adsorbed per unit mass of total sedimentary solids;
ψ = surface activity coefficient.

In most sedimentary situations, we will deal with essentially pure end-member solids such that $X \approx 1$ and $\lambda \approx 1$ so that in general:

$$a \approx 1 \quad \text{(for solids).} \tag{4-9}$$

Also, most pore water solutions (including seawater) are sufficiently dilute that we can assume that $X_{H_2O} \approx 1$, $\lambda_{H_2O} \approx 1$ (e.g., see Garrels and Christ, 1965) and therefore:

$$a \approx 1 \quad \text{(for } H_2O \text{ in pore solution).} \tag{4-10}$$

In this case, $\rho_w^* \approx 1$, so that for dissolved species, when the activity of water is one:

$$a \approx \gamma C. \tag{4-11}$$

The description of adsorption can also be simplified, but this is discussed separately in the next section devoted to adsorption.

From the above, it is obvious that we will be concerned mainly with determining activity coefficients γ for dissolved species in interstitial solution. However, it is not our goal here to go into detailed discussion of methods of activity coefficient calculation, which are covered extensively in such works as that by Robinson and Stokes (1959), or, in the case of natural waters, by Garrels and Christ (1965) and Whitfield (1975). Here we will introduce some simplifying assumptions that will enable relatively rapid calculation. First, we assume that the concentration C represents the analytical concentration or sum of free ions plus all dissolved ion pairs, complexes, etc. formed by the ion of interest. For example, in the case of sulfate ion:

$$C_{SO_4^{--}} = C_{\text{free } SO_4^{--}} + C_{NaSO_4^-} + C_{KSO_4^-} + C_{MgSO_4^0} + C_{CaSO_4^0} + \sum C_{\text{organic } SO_4^{--}}. \tag{4-12}$$

With this convention, the activity coefficient γ becomes a total activity coefficient γ_T (Berner, 1971), representing the ratio between activity and total or analytical concentration. The same convention has been used in calculating diffusion coefficients (Lasaga, 1979; see Table 3-1). Next we assume that in interstitial solutions of approximately the same salinity and composition as seawater, values of total activity coefficients for major ions are the same as those listed in Table 4-1 for average seawater at 25°C, 1 atm total pressure. This thus facilitates rapid calculation of approximate total concentration from thermodynamic activity data. If more accurate calculation is required, correction for the effects of temperature and pressure on activity coefficients in seawater can be done for a number of

TABLE 4-1

Total activity coefficients γ_T for the major ions in seawater. $T = 25°C$, $P = 1$ atm, $S = 35‰$. (Data from Whitfield, 1975, using a specific interaction model; and Pytkowicz, 1975.)

Ion	γ_T
Cl^-	0.681
Na^+	0.652
Mg^{++}	0.215
SO_4^{--}	0.121
Ca^{++}	0.201
K^+	0.618
HCO_3^-	0.50
CO_3^{--}	0.030

species using the partial molar volume and enthalpy data and procedure given by Millero (1979) for seawater. Correction for appreciable changes in salinity or major ion concentrations requires laborious calculation which can be avoided through use of standard computer programs (e.g., Truesdell and Jones, 1974, WATEQ). Computer programs are most appropriate for calculating γ_T values for salinities less than that of seawater. For salinities greater than that of seawater, other methods, such as that of Wood (1975), are more accurate.

A further simplifying assumption is that activity coefficients for dissolved solutes in sediment pore water are constant with depth and time. This is reasonable if deposition has occurred in water of roughly constant salinity and major ion composition, as is the usual case for marine sediments. Also, we will assume that K is constant with depth, or, in other words, that pressure and temperature gradients are not large enough to bring about appreciable changes in K. This is a reasonable assumption for early diagenesis where burial does not exceed a few hundred meters.

For freshwater sediments with salinities of less than about 500 mg/liter, it is relatively simple to determine values of γ_T using the equation:

$$\gamma_T = \frac{m}{m_T} \gamma^*, \tag{4-13}$$

where m_T = total molality for a given element (free ions plus ion pairs);
m = molality of the free ion;
γ^* = activity coefficient given by the Debye-Hückel limiting law.

The molality of the free ion is normally calculated from mass balance and ion-pair equilibrium expressions using the method of circular-refinement (Garrels and Christ, 1965; Stumm and Morgan, 1970; Berner, 1971). However, at salinities of <500 mg/l, for the principal ions (e.g., Na^+, K^+, Mg^{++}, Ca^{++}, Cl^-, SO_4^{--}), m/m_T is generally greater than 0.75 and for many purposes may be assumed equal to one. Values of γ^* are calculated via the Debye-Hückel expression:

$$-\log \gamma_i^* = \frac{AZ_i^2\sqrt{I}}{1 + \mathring{a}B\sqrt{I}}, \tag{4-14}$$

where A, B = constants depending only on temperature (see Garrels and Christ, 1965);
$\mathring{a}$ = ion-size parameter for ion i (see also Garrels and Christ, 1965);
Z_i = valence of ion i;
I = ionic strength.

Ionic strength is defined as:

$$I = \tfrac{1}{2}\sum m_i Z_i^2, \tag{4-15}$$

where summation is over all ionic species including i. Ionic strength is a measure of the concentration of charge in solution contributed by all ions, and constitutes the major control on γ_i^*.

For calcium and carbonate species in seawater, a separate procedure has been established for the calculation of equilibrium concentrations (e.g., Edmond and Gieskes, 1970; Takahashi, 1975; Morse and Berner, 1979). Instead of calculating activity equilibrium constants from thermodynamic data, one uses published values of directly measured concentration equilibrium constants for a given temperature, pressure, and salinity. The constants are given directly in terms of total molalities and the (operationally defined) activity of hydrogen ion. For example, for bicarbonate ion:

$$HCO_3^- \leftrightharpoons CO_3^{--} + H^+,$$

$$K' = \frac{a_{H^+} m_{T_{CO_3^{--}}}}{m_{T_{HCO_3^-}}} \tag{4-16}$$

where a_{H^+} = activity of hydrogen ion as determined in seawater by conventional electrode techniques;

m_T = total molality (free ions plus ion pairs)

K' = empirically determined equilibrium constant (a function of temperature, pressure, *and* salinity).

This procedure eliminates the necessity of using activity coefficients.

Solubility Equilibrium

A very common equilibrium reaction encountered in sediments is that between a component of a solid and its ions in solution. For example, consider the mineral barite:

$$\begin{aligned} BaSO_{4\,\text{barite}} &\leftrightharpoons Ba^{++} + SO_4^{--} \\ K &= a_{Ba^{++}} a_{SO_4^{--}}. \end{aligned} \tag{4-17}$$

Since, as stated above, we assume that barite is pure $BaSO_4$, we are left with only activities of dissolved species in the expression for the equilibrium constant (a of $BaSO_4$ in barite $= 1$). In this case, K is known as the activity solubility product, or equilibrium ion activity product. In terms of total concentration (see equation 4-6), equation (4-17) is:

$$K = \left[\frac{\gamma_{Ba^{++}} \gamma_{SO_4^{--}}}{(\rho_w^*)^2}\right] C_{Ba^{++}} C_{SO_4^{--}}. \tag{4-18}$$

It is often convenient to simplify equations like (4-18) so that we may introduce the concentration solubility product, K_c, such that

$$K_c = \left[\frac{(\rho_w^*)^2}{\gamma_{Ba^{++}}\gamma_{SO_4^{--}}}\right] K = C_{Ba^{++}} C_{SO_4^{--}}. \tag{4-19}$$

Now, suppose that a gradient in the concentration of dissolved sulfate occurs in a sediment as the result of bacterial sulfate reduction, but that solubility equilibrium with barite is maintained at all depths. If so, the gradient in $C_{Ba^{++}}$ is directly related to the gradient in $C_{SO_4^{--}}$. Equation (4-18) applies to all depths and all times, so that for constant K, γ, and ρ_w^* (the conventional assumption stated above):

$$\frac{\partial C_{Ba^{++}}}{\partial x} = -\frac{C_{Ba^{++}}}{C_{SO_4^{--}}} \frac{\partial C_{SO_4^{--}}}{\partial x}, \tag{4.20}$$

$$\frac{\partial C_{Ba^{++}}}{\partial x} = -\frac{K_c}{C_{SO_4^{--}}^2} \frac{\partial C_{SO_4^{--}}}{\partial x}. \tag{4-21}$$

These expressions along with (4-19) can be used to link the individual diagenetic equations for Ba^{++} and SO_4^{--}, and enable us to solve them for the concentration of each ion.

Once the equilibrium ion activity product is known, one can compare it to the actual ion activity product in solution to determine the state of saturation of the solution. For this we introduce the dimensionless parameter Ω such that:

$$\Omega = IAP/K = ICP/K_c, \tag{4-22}$$

where IAP = actual ion activity product (for a salt A_xB_y, $IAP = a_A^x a_B^y$);
ICP = actual ion concentration product;
K = equilibrium ion activity product (solubility product);
K_c = equilibrium ion concentration product.

Now, once the value of Ω is ascertained, we can express the state of saturation as follows:

$\Omega > 1$, solution is supersaturated.
$\Omega = 1$, solution is saturated.
$\Omega < 1$, solution is undersaturated.

This convention is a useful measure of the departure from equilibrium, and is used extensively when referring to the saturation state of seawater with respect to calcium carbonate.

Equilibrium Adsorption and Ion exchange

Sediments often contain abundant fine-grained material with grain sizes of less than one micron. Consequently, large surface areas are available for exchange of ions and molecules between the solids and the interstitial water. If the exchange involves only the surfaces of the solids, adsorption and desorption occur, and they most commonly are very fast. In fact, because of this rapidity, the processes of adsorption and ion exchange are normally treated as equilibrium reactions. This is the procedure adopted in this book. It is realized that adsorption and exchange can also be slow, but unfortunately, little is known about the rates and mechanisms, i.e., the kinetics of adsorption in natural environments. (An exception is the work of deKanel and Morse, 1978). Before proceeding to mathematical treatment in terms of diagenetic equations, it is necessary to delve into the field of surface chemistry to try to obtain a better understanding of the phenomena under discussion.

Adsorption takes place as a consequence of the attraction of ions or molecules to the surface of a solid. The attraction may result from weak, van der Waals-type forces, in which case the process is known as physical adsorption, or it may involve the formation of strong chemical bonds giving rise to chemisorption (Adamson, 1967). Chemisorption is the process of major interest to the study of sediment diagenesis, and it may involve neutral species (e.g., H_2O, H_4SiO_4, organic molecules) or ions. Increased chemisorption (henceforth referred to simply as adsorption) accompanies increases in concentration of the adsorbing species, and at equilibrium, various adsorption relations, often referred to as "isotherms," are obeyed. Two common examples are the Langmuir isotherm:

$$\bar{C} = \frac{AC}{B + C}, \tag{4-23}$$

where $\bar{C}$ = concentration of an adsorbed species in terms of mass per unit mass of total sediment solids,

C = concentration in solution,

A, B = constants for a given temperature,

and the Freundlich isotherm

$$\bar{C} = AC^{1/n}, \tag{4-24}$$

where A, n = constants for a given temperature. A special case of the Langmuir isotherm is where $B \gg C$. This results in the simple *linear isotherm* commonly used in diagenetic models:

$$\bar{C} = K'C, \tag{4-25}$$

where $K' = A/B$.

Preferential adsorption of ions results in the production of a surface charge on the adsorbing solid particles and an electrical potential associated with them. Ions producing this charge are referred to as potential-determining ions, and they may be either positive or negative. Simultaneous adsorption of positive and negative ions leads to the concept of the *zero* point of charge where total charge from adsorbed cations and anions present on the surface is zero. The zero point of charge does not, in general, correspond to equal concentrations in solution of the adsorbing cation and anion, because each may be adsorbed more strongly for a given concentration than the other. For simple salts (e.g., AgI), the usual potential-determining ions are those contained within the solid (i.e., Ag^+ and I^-). For the common oxide and silicate minerals found in sediments, the most common potential-determining ions are H^+ and OH^- or complex ions formed by the reaction of H^+ and OH^- with the mineral surfaces (Stumm and Morgan, 1970). The pH corresponding to the zero point of charge is referred to as pH_{zpc}, or, occasionally, as the isoelectric point. Some representative values of pH_{zpc} taken from the data summarized by Stumm and Morgan (1970) are shown in Table 4-2.

TABLE 4-2

Values of pH for the zero point of charge for selected substances. Note that for a typical marine sediment with pH = 7.5, all substances with $pH_{zpc} > 7.5$ will be positively charged and all those with $pH_{zpc} < 7.5$ will be negatively charged. (After Stumm and Morgan, 1970.)

Material	pH_{zpc}
α-Al_2O_3	9.1
α-$Al(OH)_3$	5.0
Fe_3O_4	6.5
γ-Fe_2O_3	6.7
$Fe(OH)_3$ (amorph.)	8.5
MgO	12.4
MnO_2	2–4.5
SiO_2	2.0
$ZrSiO_4$	5.0
Kaolinite	4.6
Montmorillonite	2.5
Albite	2.0

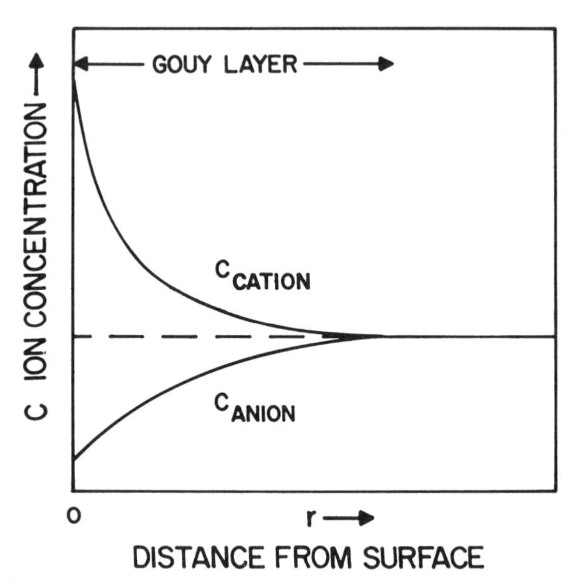

FIGURE 4-1. Distribution of counterions about a negatively charged surface according to the Gouy model. Note the excess of cations over anions within the Gouy layer. (After Van Olphen, 1977.)

As a result of the adsorption of potential-determining ions, a phenomenon known as the *electrical double layer* results. The inner part of the double layer, known as the *fixed layer*, is made up of the potential-determining ions themselves (see also clay double layers below). The outer part is referred to as the mobile or *diffuse layer* and consists of more or less freely diffusing *counterions* attracted (or repelled) by the charge set up by the potential-determining ions. The diffuse layer results solely from electrostatic effects, and thus does not exhibit the sharpness and strong specificity of the fixed layer. The counterions are represented by major ions which happen to be present in the outer solution and which readily undergo exchange for one another whenever the composition of the solution is changed. This gives rise to the process of ion (counterion) exchange. The counterions are present in a gradually increasing (or decreasing) concentration as the surface is approached. The actual distribution represents a balance between electrostatic attraction (or repulsion) and diffusive forces set up by the resulting concentration gradients, and can be calculated according to the Gouy theory (e.g., see Van Olphen, 1977). An example is shown in Figure 4-1.

Electrical double layers on sediment grains need not be produced only by the adsorption of potential-determining ions. Certain clay minerals,

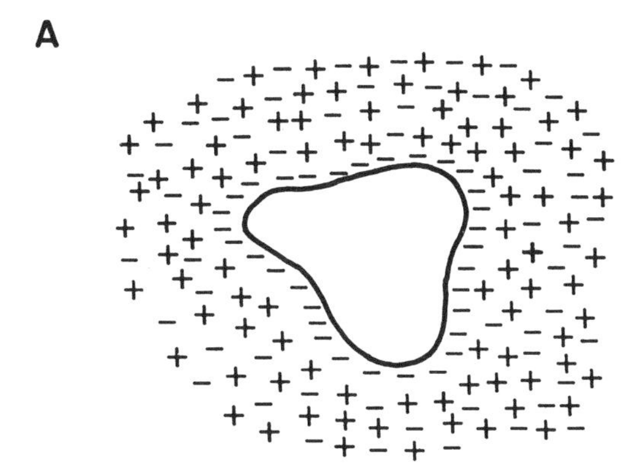

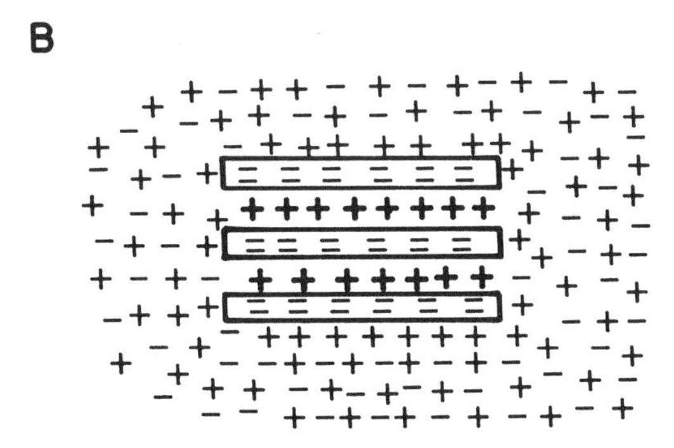

FIGURE 4-2. Two types of electrical double layers. The situation depicted in A represents adsorption, on the surface, of potential-determining ions (minus charges) with Gouy-type adsorption in the outer layer. The situation depicted in B represents interlayer adsorption of positive counterions (boldface plus-charges) to balance negative charge on clay layer (rectangles).

chiefly smectites and degraded illites, undergo ionic substitution within the alumino-silicate framework, which gives rise to a net negative surface charge. This charge is neutralized by hydrated cation adsorption between the alumino-silicate atomic layers making up each clay grain and on the flat outer portions of the grains. The cations readily exchange with ions in solution, and thus represent a special kind of counterion (see Fig. 4-2).

Because of the abundance of smectites and illite, cation exchange in fine-grained sediments is normally described in terms of clay-type double layers. However, it is likely that much exchange also comes about as a result of the adsorption of potential-determining ions. For example, iron and manganese oxides are often present as semi-amorphous, colloidal particles on larger sediment grains (including clay grains), and, because of their very high specific surface area, adsorb appreciable quantities of potential-determining ions (Jenne, 1968). In addition, clay minerals themselves are normally very fine grained and expose many broken bonds to solution where adsorption of potential-determining ions may occur. In fact, the cation exchange properties of the clay mineral kaolinite must be due to this process, since appreciable atomic substitution within the kaolinite lattice does not occur. Potential-determining ions need not be simple ions like H^+ and OH^-. A good example is provided by complex organic anions resulting from the dissociation of humic and fulvic acids. Organic anions adsorbed on fine mineral grains, including clays, may expose deprotonated carboxyl groups $RCOO^-$ to solution and thereby attract various cations from solution. Processes like this are emphasized by Schnitzer and Khan (1972), who believe that organic matter may exert a major control on the ion-exchange properties of soils and sediments. Further evidence is provided by the high cation exchange capacity of humic substances as compared to clay minerals (Kononova, 1966), and by the observations of Rosenfeld (1979) who found that NH_4^+–K^+ exchange by organic-rich marine sediments decreased considerably upon the oxidative removal of organic matter. These observations all strongly point to the need for further study of the ion-exchange properties of organic matter and fine-grained solids other than clay minerals.

Mathematical representation of equilibrium ion exchange in the most general sense is in terms of activities. In other words, for the reaction (taking cations as an example):

$$nA_{aq}^{+m} + mB_s^{+n} \leftrightarrows mB_{aq}^{+n} + nA_s^{+m},$$

$$K = \frac{a_{B_{aq}}^{m} a_{A_s}^{n}}{a_{A_{aq}}^{n} a_{B_s}^{m}}, \tag{4-26}$$

where the subscripts s and aq refer respectively to absorbed and dissolved A or B. Note that this reaction is written so as to conserve surface charge, which is appropriate when dealing with counterion exchange. In terms of concentrations and activity coefficients, using the conventions given in equations (4-6) and (4-8):

$$K = \frac{C_B^m \bar{C}_A^n}{C_A^n \bar{C}_B^m}\left(\frac{\gamma_B^m \psi_A^n}{\gamma_A^n \psi_B^m}\right). \tag{4-27}$$

In general, as shown earlier, it is possible to calculate activity coefficients for the dissolved ions, γ. However, no simple theory is available for the determination of activity coefficients for adsorbed ions. Various models have been proposed, such as that of Truesdell and Christ (1968), where regular solution theory is applied to K^+, Na^+, and NH_4^+ exchange, with H^+ on clay minerals, but more research is needed before a general model is adopted. Here, for the sake of brevity, we will simply ignore adsorption activity coefficients, thereby assuming that $\psi = 1$. This would be the correct procedure if the exchanging ions formed an ideal surface solution, and in many situations this is not a bad approximation (e.g., see Walton, 1959; or Helfferich, 1962).

A very simple relation results for the ideal exchange of two ions with the same valence. For example, the above reaction for two monovalent cations is:

$$A_{aq}^{+} + B_{s}^{+} \leftrightarrows A_{s}^{+} + B_{aq}^{+},$$

$$K = \frac{C_B \bar{C}_A}{C_A \bar{C}_B}\left(\frac{\gamma_B}{\gamma_A}\right), \tag{4-28}$$

or upon rearranging:

$$\frac{\bar{C}_B}{\bar{C}_A} = \left(\frac{K\gamma_A}{\gamma_B}\right)^{-1} \frac{C_B}{C_A}. \tag{4-29}$$

For ions of equal valence a good approximation for most solutions is that $\gamma_A = \gamma_B$. Thus, one can further simplify (4-29) to:

$$\frac{\bar{C}_B}{\bar{C}_A} = K^{-1} \frac{C_B}{C_A}. \tag{4-30}$$

This is the most simple representation of ion exchange, and shows that the ion-ratio on the solid surface is directly proportional to the same ratio in solution regardless of actual concentrations.

If the concentration of one ion is much greater than the other, even further simplification is possible. If $C_A \gg C_B$ and $\bar{C}_A \gg \bar{C}_B$, then there is little change in either C_A or $\bar{C}_A$ upon exchange of B for A. In this case equation (4-30) can be written as:

$$\bar{C}_B = K' C_B, \tag{4-31}$$

where $K' = K^{-1}\bar{C}_A/C_A \approx$ constant. Note that this expression is identical in form to that for simple linear adsorption (equation 4-25).

If two ions of different valence exchange for one another, we have a considerably more complicated situation. For example, consider mono- and divalent cations undergoing ideal exchange:

$$2A_{aq}^{+} + B_{s}^{++} \leftrightarrows B_{aq}^{++} + 2A_{s}^{+}$$

$$K = \frac{C_B \bar{C}_A^2}{C_A^2 \bar{C}_B}\left(\frac{\gamma_B}{\gamma_A^2}\right), \tag{4-32}$$

or rearranging:

$$\frac{\bar{C}_A^2}{\bar{C}_B} = \left(\frac{K\gamma_A^2}{\gamma_B}\right)\frac{C_A^2}{C_B}. \tag{4-33}$$

This is the Donnan equilibrium formulation for this reaction (Helfferich, 1962). In general, for different valence ions, $\gamma_B \neq \gamma_A$, and even if γ values were equal, the ratio γ_B/γ_A^2 does not reduce to unity. More importantly, the ratio of A to B on the solid is not simply related to the same ratio in solution, but instead depends upon the absolute concentration of the ions. This can be seen by rewriting (4-33) as:

$$\frac{\bar{C}_A}{\bar{C}_B} = \left(\frac{K\gamma_A^2 C_A}{\gamma_B \bar{C}_A}\right)\frac{C_A}{C_B}. \tag{4-34}$$

Now, since under these conditions $C_A/\bar{C}_A$ is not a constant, it is possible to change the ratio $\bar{C}_A/\bar{C}_B$ by lowering the concentration of dissolved A and B, even while maintaining a constant ratio of C_A/C_B. This expresses the well-known observation that upon dilution, a greater proportion of the higher valent ion is taken up by an ion exchanger (e.g., Sayles and Mangelsdorf, 1977). As pointed out by Sayles and Mangelsdorf, and by Murthy and Ferrell (1972), this fact has often been neglected by individuals who, when measuring ion exchange between clays and seawater, wash out, and thereby dilute, the interstitial seawater, in order to determine adsorbed cations. This results in an inadvertant enrichment, during the washing, of Mg^{++} on the clays.

Inclusion of Equilibrium Processes in Diagenetic Equations

To demonstrate how equilibrium processes are included in diagenetic equations, two simple but representative examples have been chosen. They are solubility equilibrium of a binary salt whose ions are of the same valence, and simple linear adsorption. These examples are sufficient to demonstrate the basic principles involved while not overburdening the reader with unnecessary mathematical complexity. For a more general treatment of adsorption in diagenetic equations, see Berner (1976) and Schink and Guinasso (1978a).

For the case of solubility equilibrium of a binary salt, AB, whose ions are not involved in any other equilibrium reactions, we can write four expressions. The first are diagenetic equations for dissolved A and B. Adopting the convention $\hat{C} = \phi C$, from the general diagenetic equation

(2-13), modified to account separately for bioturbation and molecular diffusion (equation 3-74), we obtain:

$$\frac{\partial(\phi C_A)}{\partial t} = \frac{\partial\left[D_B \dfrac{\partial(\phi C_A)}{\partial x} + \phi(D_{s_A} + D_I)\dfrac{\partial C_A}{\partial x}\right]}{\partial x} - \frac{\partial(\phi v C_A)}{\partial x} + \phi R_{A_{sol}} + \phi \sum R'_A, \tag{4-35}$$

$$\frac{\partial(\phi C_B)}{\partial t} = \frac{\partial\left[D_B \dfrac{\partial(\phi C_B)}{\partial x} + \phi(D_{s_B} + D_I)\dfrac{\partial C_B}{\partial x}\right]}{\partial x} - \frac{\partial(\phi v C_B)}{\partial x} + \phi R_{B_{sol}} + \phi \sum R'_B, \tag{4-36}$$

where R_{sol} = rate of change of the concentration of dissolved A or B due to dissolution or precipitation of salt AB in mass per unit volume of pore water per unit time;

R' = all non-equilibrium slow reactions affecting A and B;

and as defined elsewhere,

ϕ = porosity;

C = concentration of dissolved A or B in mass per unit volume of pore water;

t = time;

x = depth;

D_B = solid bioduffusion coefficient;

D_I = irrigation biodiffusion coefficient;

D_s = molecular diffusion coefficient;

v = velocity of pore water burial.

To these two expressions are added the stoichiometric expression appropriate to a solid of composition AB:

$$R_{A_{sol}} = R_{B_{sol}}. \tag{4-37}$$

Finally, we have the equilibrium expression, which from equation (4-19) (ignoring activity coefficients) is:

$$K = C_A C_B. \tag{4-38}$$

Assuming that $D_{s_A} = D_{s_B} = D_s$, and combining equations (4-35), (4-36), (4-37), and (4-38), we obtain:

$$\frac{\partial[\phi(C_A - K/C_A)]}{\partial t} = \frac{\partial\left\{D_B \frac{\partial[\phi(C_A - K/C_A)]}{\partial x} + \phi(D_s + D_I)\frac{\partial(C_A - K/C_A)}{\partial x}\right\}}{\partial x} - \frac{\partial[\phi v(C_A - K/C_A)]}{\partial x} + \phi[\sum R'_A - \sum R'_B]. \tag{4-39}$$

Terms involving the space and time derivatives of ϕ can be largely eliminated from this expression, through use of the analogous diagenetic equation for total water (see Chapter 3). If the chemical reactions do not involve appreciable water production or consumption,

$$\frac{\partial \phi}{\partial t} = \frac{\partial\left(D_B \frac{\partial \phi}{\partial x}\right)}{\partial x} - \frac{\partial(\phi v)}{\partial x}. \tag{4-40}$$

Multiplying both sides of (4-40) by $(C_A - K/C_A)$ and subtracting from (4-39), we obtain:

$$\frac{\partial(C_A - K/C_A)}{\partial t} = \frac{1}{\phi}\frac{\partial\left[\phi(D_B + D_I + D_s)\frac{\partial(C_A - K/C_A)}{\partial x}\right]}{\partial x} - \left(v - \frac{D_B}{\phi}\frac{\partial \phi}{\partial x}\right)\frac{\partial(C_A - K/C_A)}{\partial x} + \sum R'_A - \sum R'_B. \tag{4-41}$$

This expression shows the effects of simple solubility equilibrium that can be incorporated directly into a diagenetic equation for species A, but it produces a complicated non-linear concentration dependency of A in the resulting expression.

Inclusion of simple linear adsorption (equation 4-25) in diagenetic equations proceeds along somewhat similar lines, except that we now deal with only one species i which is present in both the dissolved and adsorbed state. The appropriate diagenetic equation for dissolved i (from equation

3-74 is:

$$\frac{\partial(\phi C)}{\partial t} = \frac{\partial\left[D_B \frac{\partial(\phi C)}{\partial x} + \phi(D_s + D_I)\frac{\partial C}{\partial x}\right]}{\partial x} - \frac{\partial(\phi v C)}{\partial x} + \phi R_{ads} + \phi \sum R', \tag{4-42}$$

where R_{ads} = rate of change of dissolved i due to equilibrium adsorption or desorption in mass per unit volume of pore water per unit time,

R' = all other slow (irreversible) reactions affecting i,

and all other parameters are defined as above. For adsorbed i, we introduce the notation:

$$\bar{C} = C_s \tag{4-43}$$

into the general diagenetic equation for solids (equation 3-73):

$$\frac{\partial[(1-\phi)\bar{\rho}_s \bar{C}]}{\partial t} = \frac{\partial\left\{D_B \frac{\partial[(1-\phi)\bar{\rho}_s \bar{C}]}{\partial x}\right\}}{\partial x} - \frac{\partial[(1-\phi)\bar{\rho}_s \omega \bar{C}]}{\partial x} + (1-\phi)\bar{\rho}_s \bar{R}_{ads} + (1-\phi)\bar{\rho}_s \sum \bar{R}', \tag{4-44}$$

where $\bar{C}$ = concentration of adsorbed i in mass per unit mass of total solids;

$\bar{R}_{ads}$ = rate of change of adsorbed i due to equilibrium adsorption or desorption in mass per unit mass of total solids per unit time;

$\bar{R}'$ = all non-equilibrium slow reactions affecting adsorbed i;

$\bar{\rho}_s$ = average density of total solids;

ω = rate of depositional burial of solids.

In addition to the diagenetic equations, we have the mass balance expression:

$$\bar{R}_{ads} = \frac{-\phi}{(1-\phi)\bar{\rho}_s} R_{ads} \tag{4-45}$$

and the equilibrium expression for simple linear adsorption (4-25):

$$\bar{C} = K'C, \tag{4-46}$$

where K' = adsorption constant.

To simplify our treatment, we will assume that the adsorptive properties do not change with depth or time, in other words, that $K' =$ constant. If so, then:

$$\frac{\partial \bar{C}}{\partial t} = K' \frac{\partial C}{\partial t}, \tag{4-47}$$

$$\frac{\partial \bar{C}}{\partial x} = K' \frac{\partial C}{\partial x}. \tag{4-48}$$

To eliminate porosity dependence from the final expression as much as possible, we may use the analogous diagenetic equations for pore water and total solids:

$$\frac{\partial \phi}{\partial t} = \frac{\partial \left[D_B \frac{\partial \phi}{\partial x} \right]}{\partial x} - \frac{\partial (\phi v)}{\partial x}, \tag{4-49}$$

$$\frac{\partial [(1 - \phi)\bar{\rho}_s]}{\partial t} = \frac{\partial \left\{ D_B \frac{\partial [(1 - \phi)\bar{\rho}_s]}{\partial x} \right\}}{\partial x} - \frac{\partial [(1 - \phi)\bar{\rho}_s \omega]}{\partial x}. \tag{4-50}$$

Multiplying both sides of (4-49) by C, both sides of (4-50) by $\bar{C}$, and combining the resultant expressions with (4-42) and (4-44) through (4-48), we obtain:

$$\begin{aligned} \frac{\partial C}{\partial t} = {} & \left[\frac{1}{\phi(1 + K)} \right] \frac{\partial \left\{ \phi[(1 + K)D_B + D_I + D_s] \frac{\partial C}{\partial x} \right\}}{\partial x} \\ & - \left\{ \frac{\phi(v + K\omega) - [(1 + K)D_B] \frac{\partial \phi}{\partial x} - \phi D_B \frac{\partial K}{\partial x}}{\phi(1 + K)} \right\} \frac{\partial C}{\partial x} \\ & + \frac{\sum R'}{1 + K} + \frac{(1 - \phi)\bar{\rho}_s \sum \bar{R}'}{\phi(1 + K)} \end{aligned} \tag{4-51}$$

where $K = [\bar{\rho}_s(1 - \phi)/\phi]K'$.

This equation is similar to that derived by Schink and Guinasso (1978a), except that here bioturbation of total solids and total pore water is not neglected as was done by these authors. One can see that the effects of adsorption, with the simple linear model, are readily incorporated into the diagenetic equation in terms of K. If porosity and average density and adsorptive properties of solids are constant with depth, and there is no externally impressed water flow, then equation (4-51) can be considerably simplified. Under these conditions:

$$\frac{\partial \bar{\rho}_s}{\partial x} = 0, \frac{\partial \phi}{\partial x} = 0,$$

$$\frac{\partial K}{\partial x} = 0, \omega = v.$$

Substituting in (4-51), we obtain:

$$\frac{\partial C}{\partial t} = \frac{\partial \left(D_B \frac{\partial C}{\partial x} \right)}{\partial x} + \left(\frac{1}{1 + K} \right) \frac{\partial \left[(D_I + D_s) \frac{\partial C}{\partial x} \right]}{\partial x} - \omega \frac{\partial C}{\partial x}$$
$$+ \left(\frac{1}{1 + K} \right) \left[\sum \bar{R}' + \frac{(1 - \phi)\bar{\rho}_s}{\phi} \sum \bar{R}' \right]. \tag{4-52}$$

Equation (4-52) shows that the effect of increased adsorption (increasing K) is to decrease the importance of irrigation, molecular diffusion, and most chemical reaction relative to solid biodiffusion and advective burial. (If surface reactions are important, the effect of increased adsorption on chemical reaction depends on the functional dependence of $\sum \bar{R}'$ upon $\bar{C}$. In some situations, such as radioactive decay both on surfaces and in solution, there is no net effect of adsorption on chemical reaction—see Berner, 1976.) As a limit where K becomes very large, the equation in general reduces simply to that for biodiffusion and burial. This result can be visualized (Schink and Guinasso, 1978a) as the enhanced transport of a dissolved component i, by means of extensive adsorption on solids in regions of production of i, bioturbational migration of the solids, and release from the solids to pore water in regions of consumption or no production of i. Without adsorption plus bioturbation, dissolved i could only migrate via molecular diffusion which, under these circumstances, would be much slower.

Equation (4-52) can be further simplified for sediments below the zone of bioturbation where $D_B = 0$, $D_I = 0$. Since we have assumed that $\partial \phi / \partial x = 0$, it is also reasonable to assume that $\partial D_S / \partial x = 0$ (see

"Diffusion" section of Chapter 3). Under these conditions, equation (4-52) reduces to:

$$\frac{\partial C}{\partial t} = \left(\frac{D_s}{1 + K}\right)\frac{\partial^2 C}{\partial x^2} - \omega\frac{\partial C}{\partial x} + \left(\frac{1}{1 + K}\right)\left[\sum R' + \left(\frac{1 - \phi}{\phi}\right)\bar{\rho}_s \sum \bar{R}'\right]. \tag{4-53}$$

Note here that the effect of increased adsorption is to increase the relative importance of depositional burial (and in some instances surface reactions). The burial term is usually neglected in diagenetic modeling of deep-sea sediments due to low ω. One should be careful, however, to determine the magnitude of K before automatically eliminating the burial term from diagenetic equations; for conditions typical of deep-sea sediments, if $K > 100$, burial cannot be neglected (Berner, 1976).

If there are no slow diagenetic reactions ($\sum R' = 0; \sum \bar{R}' = 0$), and burial is negligible, equation (4-52) reduces to:

$$\frac{\partial C}{\partial t} = \left(\frac{D_s}{1 + K}\right)\frac{\partial^2 C}{\partial x^2}. \tag{4-54}$$

This is the Fick's Second Law expression for dissolved species undergoing diffusion plus equilibrium adsorption. It has been used extensively to describe the diffusion of radionuclides into sediments as a result of sudden deposition at the sediment-water interface accompanying radioactive waste disposal (e.g., Duursma and Hoede, 1967). It has also been used (e.g., by Li and Gregory, 1974) to determine values of K from lab diffusion and adsorption experiments. Treatment of diffusion in diagenetic equations in terms of a lowered diffusion coefficient, $D' = D_s/(1 + K)$, due to adsorption, has often been done by others, but is correct only for the many assumptions leading to equation (4-54). In other words, in general, D' cannot be used in place of D_s, because $(1 + K)$ appears in other terms of the diagenetic equation, and besides, the use of constant $(1 + K)$ rests upon the assumption of simple linear equilibrium adsorption and negligible porosity change with depth which may, in many situations, be incorrect.

Values of K derived from laboratory adsorption and diffusion experiments for a variety of dissolved species and sediments are listed in Table 4-3. It should be noted that in most cases checks of the assumption of simple linear, reversible adsorption were *not* made. Thus, many of the results in Table 4-3 should be considered only as rough, order-of-magnitude values.

TABLE 4-3

Values of K (dimensionless, porosity-dependent adsorption constant) for near-surface (<2 m burial depth) fine-grained marine clay muds (assuming simple linear adsorption). Diffusion method refers to use of equation (4-54) and calculated values of D_s.

Ion	K	ϕ	*Method and reference*
Na^+	0.3	0.71	Diffusion (1)
Na^+	0.3	0.71	Adsorption (1)
Ca^{++}	1.8	0.71	Diffusion (1)
Ca^{++}	1.4	0.71	Adsorption (1)
NH_4^+	1.6	0.63	Adsorption (2)
NH_4^+	1.3	0.75	Adsorption (2)
PO_4^{---}	2.0	0.71	Diffusion (3)
PO_4^{---}	(1.2–2.5)	0.71	Adsorption (3)
Sr^{++}	4–7	0.64	Diffusion (4)
Sr^{++}	1–10	0.64	Adsorption (4)
Cs^+	750	0.64	Diffusion (4)
Cs^+	2,500–5,000	0.64	Adsorption (4)
Ra^{++}	5,000–10,000	0.7	Adsorption and Diagenetic Modeling (5)

(1) Li and Gregory (1974).
(2) Rosenfeld (1979).
(3) Krom and Berner (1980).
(4) Duursma and Bosch (1970).
(5) Cochran (1980).

Homogeneous Reactions (Including Radioactive Decay)

Reactions which occur wholly within a single phase are referred to as being homogeneous. In sediments, homogeneous *chemical* reactions are essentially all restricted to the interstitial water. Such reactions, if they do not involve oxidation or reduction, are fast and can be considered as being at equilibrium, a common example being the dissociation of carbonic acid to $HCO_3^- + H^+$. By contrast, several homogeneous redox reactions can be slow, in fact very slow in the case of the non-biological reduction of sulfate to sulfide by dissolved organic compounds. In most cases, however, such redox reactions are mediated by micro-organisms, and thus, cannot be classified as being homogeneous. Because of their importance to diagenesis, they are discussed in a separate section (see below).

Homogeneous reactions can be classified according to the rate law, which they follow. The rate law is simply an empirical expression which shows how the rate of reaction depends upon temperature and the concentrations of reactants. In general, it cannot be deduced from the stoichiometry of the overall reaction because the rate law reflects the slowest step in a series of complex elementary reactions which together make up the overall reaction. In other words, an overall reaction such as $A \rightarrow B$ cannot be assumed to be first order with respect to A, even though it is written as a unimolecular process. Some simple examples of rate laws, for a single reactant are:

Zero order

$$\frac{dC}{dt} = -k_0, \tag{4-55}$$

where c = concentration;
t = time;
k_0 = zero order rate constant.

First order

$$\frac{dC}{dt} = -k_1 C, \tag{4-56}$$

where k_1 = first order rate constant.

Second Order

$$\frac{dC}{dt} = -k_2 C^2, \tag{4-57}$$

where k_2 = second order rate constant.

For two or more reactants, complex expressions may result (e.g., see Laidler, 1965). A simple example is the second order reaction:

$$\frac{dC_A}{dt} = -k_2 C_A C_B, \tag{4-58}$$

where A and B refer to two different reactants.

The effect of temperature on rate constants, over the small temperature changes found during early diagenesis, can usually be expressed in terms of the expression:

$$\frac{\partial \ln k}{\partial T} = \frac{U}{RT^2}, \tag{4-59}$$

where k = rate constant;

U = "activation energy" which, strictly speaking, is only an empirical temperature coefficient;

T = absolute temperature;

R = rate constant.

For a detailed discussion of reaction rates and temperature, the reader is referred to standard works on rate theory such as Laidler (1965).

The greatest use of homogeneous rate laws in the discussion of diagenesis is not for chemical reactions, but rather for the process of radioactive decay. All radioactive decay reactions are first order and are independent of temperature, pressure, or the chemical nature of the substance in which the decay occurs. In other words,

$$\frac{\partial \hat{C}}{\partial t} = -\lambda \hat{C}, \tag{4-60}$$

where λ is the (first order) decay constant and its value is independent of whether $\hat{C}$ refers to concentration in a mineral, a complex organic molecule, or in solution. Because of this great simplification, it is possible to include radioactive decay directly in a diagenetic equation for a radionuclide, once the value of λ is known. (The usual procedure is to calculate λ from the half-life, τ, via the equation, $\lambda = 0.693/\tau$).

Because values of λ are well known, one can use the diagenetic equation for a radionuclide to "date" the sediment, and thereby determine values of ω. However, many implicit assumptions must be made. First of all, this is only practicable for solids because a dissolved radionuclide can undergo redistribution via molecular diffusion. For a solid (not adsorbed) radionuclide, one can write the general expression analogous to equation (3-73):

$$\frac{\partial[(1-\phi)\bar{\rho}_s C_s]}{\partial t} = \frac{\partial\left\{D_B \dfrac{\partial[(1-\phi)\bar{\rho}_s C_s]}{\partial x}\right\}}{\partial x} - \frac{\partial[(1-\phi)\bar{\rho}_s \omega C_s]}{\partial x} - (1-\phi)\bar{\rho}_s \lambda C_s + (1-\phi)\bar{\rho}_s \sum R'_s, \tag{4-61}$$

where C_s = concentration of radionuclide in atoms (usually expressed as decays per minute) per unit mass of total solids;

R'_s = all reactions affecting the nuclide other than decay (which includes production from the decay of a parent isotope).

Now, if, and only if:

1. There are no appreciable depth variations in porosity and density (i.e., $\partial\phi/\partial x = 0$; $\partial\bar{\rho}_s/\partial x = 0$),
2. Bioturbation is negligible ($D_B = 0$),
3. Radioactive production and other reactions are unimportant,
4. Steady state is present,

then equation (4-61) reduces to:

$$\omega \frac{\partial C_s}{\partial x} - \lambda C_s = 0, \tag{4-62}$$

which for the boundary condition, $x = 0$, $C_s = C_{s0}$, yields the solution:

$$\ln (C_s/C_{s0}) = -(\lambda/\omega)x. \tag{4-63}$$

Thus, under these *stringent conditions*, a linear plot of $\ln (C_s/C_{s0})$ vs x yields the value of ω. This is the procedure followed, and the assumptions made during most determinations of the rate of sedimention. Radioactive production, if present, is accounted for by subtracting total radioactivity from that in equilibrium with the parent isotope. This "excess" concentration of the radionuclide is what is expressed as C_s.

In many sedimentary situations, one cannot make all these assumptions. This is especially true for near-surface sediments where bioturbation is important. A good example is provided by the distribution of excess ^{210}Pb in deep-sea sediments (Nozaki et al. 1977). Here measurable excesses of this isotope are found associated with particles at depths of over 5 cm. Since the half-life of ^{210}Pb is 22 years, at the overall sedimentation rate, (determined using ^{14}C) of 0.003 cm per year, one would not expect to find measurable excess ^{210}Pb below about 0.3 cm. Its occurrence must be due to bioturbation and, on this basis, Nozaki et al. calculated D_B from known values of ω and λ using the equation:

$$D_B \frac{\partial^2 C_s}{\partial x^2} - \omega \frac{\partial C_s}{\partial x} - \lambda C_s + P = 0, \tag{4-64}$$

which is equation (4-61) for constant porosity and density, constant D_B, constant production rate P, and steady state. Thus, radionuclides can be used to determine biodiffusion coefficients (see bioturbation section of Chapter 3 for further discussion) as well as rates of sedimentation.

The depth distribution of a *dissolved* radionuclide can also be described via diagenetic equations. An excellent example is provided by the work of Cochran (1980) on ^{226}Ra. (see Chapter 7 for a more detailed discussion.) Cochran modeled ^{226}Ra in terms of adsorption, biodiffusion of particles, molecular diffusion of dissolved ^{226}Ra, deposition, radioactive decay (both on surfaces and in solution), and production from particle-bound ^{230}Th. By solving these equations and applying the results to measured pore water data, he showed that they provide a good representation of the diagenetic factors affecting a soluble radionuclide (^{226}Ra) in a deep-sea sediment.

Microbial (Metabolic) Reactions

Under this category are included all diagenetic chemical reactions which are mediated by micro-organisms. Larger organisms also contribute to the total biological activity in a sediment (for example, during bioturbation), but their role in the turnover of chemical compounds is usually minor, as compared to the micro-organisms, especially the bacteria. A simple explanation for this is that bacteria, whose cell size is on the order of a few micrometers, expose a much greater surface area, per unit mass of protoplasm, to the interstitial water than do snails, clams, worms, and the like. This allows for much more rapid exchange of dissolved substances across the outer cell membrane: in other words, more rapid metabolism. Meiofauna, i.e., metazoans such as nematodes and forams. whose body size is less than 1 mm, can sometimes represent an appreciable fraction of the total sediment metabolism (Lasserre, 1976). However, meiofauna are important only in the top few centimeters of sediment, and below depths of a few tens of centimeters all larger organisms, including meiofauna, are absent. At these depths, only bacteria can account for metabolic reactions (e.g., Yingst, 1978).

A major reason why higher organisms disappear with depth is that most sediments become anoxic below the top few centimeters or few tens of centimeters and, in order to survive, the higher organisms, which all require oxygen, must maintain either constant or occasional contact with the overlying oxygenated water. As depth increases, this contact becomes more and more difficult. By contrast, many of the bacteria living in sediments are anaerobic, i.e., their metabolism does not involve 0_2. In fact, some important anaerobic bacteria, such as those that reduce sulfate, are killed by dissolved 0_2.

The processes of metabolism can be subdivided into two categories, anabolic (assimilation) processes, and catabolic (dissimilation) processes. Anabolic processes include all those which lead to the synthesis of protoplasm, hard parts, and so on, whereas catabolic processes involve the breakdown of already synthesized organic molecules to simpler molecules or inorganic species. The energy required for anabolism is furnished by catabolism. In terms of the rate and degree of turnover of materials in sediments, catabolism is far more important. In fact, the overall net result of diagenesis is the destruction of organic compounds, i.e., catabolism. In this book we will discuss only catabolic or decomposition processes, so that when we speak of bacterial sulfate reduction, for example, we will mean only dissimilatory reduction of sulfate to H_2S and not the assimilatory reduction of sulfate to form sulfur-bearing proteins within the cell.

Catabolic processes generally follow a definite succession in sediments depending upon the nature of the oxidizing agent (energy source). The first step is attack by dissolved O_2. Many different aerobic organisms large and small, use dissolved oxygen from the overlying water to destroy freshly deposited organic compounds via overall reactions of the type:

$$CH_2O + O_2 \rightarrow CO_2 + H_2O,$$

where CH_2O represents the organic compounds. This is an efficient process and most of the remains of dead plants and animals are destroyed by it (e.g., Menzel, 1974). If dissolved O_2 becomes sufficiently depleted by the aerobes due to burial or restricted renewal of oxygenated water, further organic decomposition continues by use of the dissolved oxygen in nitrate ion (NO_3^-) as an oxidizing agent. This occurs when the concentration of O_2 is lowered to less than $\sim 5\%$ of its value for fully aerated water (Thimann, 1963; Devol, 1978). The overall reaction, known as "denitrification," is:

$$5CH_2O + 4NO_3^- \rightarrow 2N_2 + 4HCO_3^- + CO_2 + 3H_2O.$$

Upon complete removal of dissolved O_2, strictly anaerobic micro-organisms become important. They use oxygen combined in iron and manganese oxides, in sulfate, and in organic matter as sources of energy. The processes of decomposition involve several intermediate steps mediated by different micro-organisms, but overall reactions can be written as we have done above for oxygen and nitrate reduction. They are:

$$CH_2O + 2MnO_2 + 3CO_2 + H_2O \rightarrow 2Mn^{++} + 4HCO_3^-,$$

$$CH_2O + 4Fe(OH)_3 + 7CO_2 \rightarrow 4Fe^{++} + 8HCO_3^- + 3H_2O,$$

$$2CH_2O + SO_4^{--} \rightarrow H_2S + 2HCO_3^-,$$

$$\begin{cases} 2CH_2O + 2H_2O \rightarrow 2CO_2 + 8(H) \text{ followed by} \\ 8(H) + CO_2 \rightarrow CH_4 + 2H_2O. \end{cases}$$

The last two reactions represent steps in the production of biogenic methane. Although methane formation is often referred to as carbon dioxide reduction (Claypool and Kaplan, 1974), it requires a reducing agent (here generalized as (H)) which must be derived from organic matter.

Thus, the overall process of methane formation is simply the disproportionation of organic matter to carbon dioxide and methane (sum of the last two reactions).

These reactions generally occur in the sequence listed above. Thus, in a typical near-shore marine sediment, we find aerobic decomposition at and above the sediment-water interface, nitrate reduction in the top few centimeters, sulfate reduction over the next meter or so, and methane production below the depth where sulfate disappears. The exact position in the sequence of iron and manganese oxide reduction is not well known, but recent work by Froelich et al. (1979) on deep-sea sediments suggests that the order given above is correct. The zone of sulfate reduction is much thicker than that for nitrate in near-shore sediments, simply because sulfate is $\sim$1,000 times more abundant in seawater than nitrate, and thus it takes much more time for it to be completely consumed. Because of its prominence during early diagenesis, much attention will be devoted to sulfate reduction when discussing specific sediments later in this book.

The reason why such a succession occurs is generally explained in terms of metabolic free energy yield for each reaction (e.g., Stumm and Morgan, 1970; Claypool and Kaplan, 1974; Froelich et al., 1979). It is assumed that the greater the energy yield of a microbial process, the greater the likelihood that it will dominate over other thermodynamically possible competing reactions. Reactions will then succeed one another, in the order of their free energy yield, as each oxidizing agent is successively exhausted. Calculated free energy changes for both products and reactants at standard state, are listed for the above reactions in Table 4-4, and, as can be seen, there is a consistent decrease in energy yield going down the list. This tends to corroborate the free energy-competition hypothesis.

TABLE 4-4

Standard state free energy changes for some bacterial reactions. (Data from Latimer, 1952; and Berner, 1971.) Data for CH_2O and MnO_2, are for sucrose and fine-grained birnessite, respectively.

Reaction	*ΔG° (kJ mol^{-1} of CH_2O)*
$CH_2O + O_2 \rightarrow CO_2 + H_2O$	−475
$5CH_2O + 4NO_3^- \rightarrow 2N_2 + 4HCO_3^- + CO_2 + 3H_2O$	−448
$CH_2O + 3CO_2 + H_2O + 2MnO_2 \rightarrow 2Mn^{++} + 4HCO_3^-$	−349
$CH_2O + 7CO_2 + 4Fe(OH)_3 \rightarrow 4Fe^{++} + 8HCO_3^- + 3H_2O$	−114
$2CH_2O + SO_4^{--} \rightarrow H_2S + 2HCO_3^-$	−77
$2CH_2O \rightarrow CH_4 + CO_2$	−58

The mechanisms of metabolic reactions are extremely complex. The reactions all involve enzymes, or natural biological catalysts, and the study of them falls into the field of enzyme kinetics, which is really a part of biochemistry. In order to avoid undue complexity, we will be concerned in this book only with the reactants and products entering and leaving the cell, and not with the complicated biochemistry occurring within the cell. Even this approach is difficult when we are confronted with natural sediments. In sediments, we must deal with many different interacting micro-organisms and a whole series of simultaneous reactions, with the product of one being the reactant or the inhibitor (poison) of one or more others. Generally, we do not know the nature of the intermediate organic molecules transferred from one group of bacteria to another, and as a result, we are generally forced to use overall reactions such as those written above. In fact, we do not have a very good idea of the nature of the complex starting materials in organic decomposition, the biopolymers (which are represented above simply as CH_2O), and we know even less about the relatively non-degradable complex molecules that may form abiologically during decomposition, the so-called "geopolymers" (e.g., see Welte, 1973). Nevertheless, a first-attempt approach to the kinetics of diagenetic microbial reactions will be made here, based by analogy on laboratory studies of enzymatic reactions.

The basic expression used to describe enzymatic reactions in the laboratory is the Michaelis-Menten equation (Dixon and Webb, 1964; Cornish-Bowden, 1976):

$$\frac{dC}{dt} = \frac{-R_{max}C}{K_m + C}, \tag{4-65}$$

where C = concentration in solution of a metabolic reactant, e.g., a dissolved organic compound, SO_4, O_2, etc.;

R_{max} = maximum rate of reaction where enzyme saturation occurs;

K_m = Michaelis constant.

This equation applies to several different kinds of enzymatic reactions of varying degrees of complexity, and its form can be derived from a number of different sets of initial assumptions. In this case, the characteristic constants R_{max} and K_m may represent a whole host of different combinations of individual rate-constants (Cornish-Bowden, 1976).

Most of the bacterial reactions with which we will be concerned involve two reactants. In this case, the rate of change of one reactant depends upon the concentration of the other, and this dependence is not expressed in the simple Michaelis-Menten equation above. However, under many circumstances the more complex expressions for two reactants reduce to the same form as this equation. For example, for the case where an ordered ternary complex ABX is formed between reactants A and B and an enzyme X, and reaction products involve little back reaction (already a simplified situation), an appropriate expression (Cornish-Bowden, p. 83) is:

$$\frac{dC_A}{dt} = \frac{-R_{max}C_AC_B}{K_i + K_m^BC_A + K_m^AC_B + C_AC_B}, \tag{4-66}$$

where C_A, C_B = concentration of A and B;
K_m^A, K_m^B = Michaelis constants for A and B reacting alone;
K_i = complex "inhibition" constant.

If B is present in much greater abundance than A, it is not appreciably consumed and, as a result, its concentration can be assumed to be constant during reaction with A. In this case, equation (4-66) can be rewritten as:

$$\frac{dC_A}{dt} = \frac{-R'_{max}C_A}{K'_m + C_A}, \tag{4-67}$$

where $R'_{max} = R_{max}C_B/(K_m^B + C_B)$;

$$K'_m = \frac{K_i + K_m^AC_B}{K_m^B + C_B}.$$

Note that, under these conditions, the form of the Michaelis-Menten equation is present. In fact, if C_B is sufficiently high that $C_B \gg K_m^B$ and $C_BK_m^A \gg K_i$ (in other words, B saturates the enzyme), then:

$$R'_{max} = R_{max}$$
$$K'_m = K_m^A.$$

The common use of the Michaelis-Menten equation for describing natural metabolic reactions involving two reactants is based on the assumption that one reactant is much more abundant and saturating.

Ideally, if we knew the nature of each micro-organism involved in each step in the process of organic matter degradation in sediments, and we knew the nature of the organic (and inorganic) compounds involved, we could determine R_{max} and K_m values in the laboratory for each reaction and thereby deduce the overall kinetics of organic matter decomposition. However, we are far from such knowledge. As a result, we are forced to make crude assumptions which can still be tested with appropriate sediment data. The first assumption made here is that the *overall* process of bacterial decomposition of complex organic molecules (whether biopolymers or geopolymers) to simple inorganic molecules and methane follows Michaelis-Menten type equations representing different fractions of the total decomposable material. In other words:

$$\frac{dG_T}{dt} = -\sum_{n=1}^{i}\left(\frac{R_{max}G_i}{K_{m_i} + G_i}\right), \tag{4-70}$$

where G_i = concentration of complex insoluble organic matter of type i (expressed as mass of organic carbon per unit mass of total solids) which can be decomposed to CO_2, H_2O, and/or CH_4 via a given bacterial process or set of processes;

$G_T = \sum G_i$ = total decomposable organic matter in the sediment (expressed as organic carbon).

The materials represented by G_i are assumed to be of sufficiently high molecular weight that they are in an insoluble form and therefore do not undergo appreciable molecular diffusion.

Equation (4-70) by itself is relatively intractable because of a general lack of knowledge of values for K_m and R_{max} and of what actually constitutes G_i. For the purposes of the present work, we further simplify the matter by making two additional assumptions. These are that only a few basic groups of substances are involved, and that $K_m \gg G$ for each group. In other words, each group decomposes according to first order kinetics. For example, assume that the total decomposable organic matter in sediments can be divided into three groups whose concentrations are denoted as G_1, G_2, and G_3. If also $K_m \gg G$ for each group, then equation (4-70) reduces to:

$$G_T = G_1 + G_2 + G_3, \tag{4-71}$$

$$\frac{dG_T}{dt} = -(k_1G_1 + k_2G_2 + k_3G_3), \tag{4-72}$$

where $k = R_{max}/K_m$.

Now, if $k_1 > k_2 > k_3$ and $G_1 \approx G_2 \approx G_3$, substance 1 will be removed faster than substance 2, which will, in turn, be removed faster than substance 3. In a given sediment, this gives rise to successive exhaustion of the three bacterial substrate types, and consequent deceleration of reaction rates with time. By contrast, in another sediment, if at the time of burial $G_3 > G_2 = G_1 = 0$, then the rates will be slower to start with, since substances of group 3 are attacked first.

The value of G_i, since it refers to a group of organic compounds, is independent of the bacterial process bringing about decomposition, whereas the value of k_i can vary with the process. For example, a set of carbohydrates will be decomposed aerobically at different rates than they would be anaerobically by sulfate reduction. Thus k_i is a function of both the nature of the decomposing material and the metabolic process under consideration. Putting it more simply, G_i expresses the amount of the organic compounds, while k_i expresses their reactivity. In general, k_i varies much more from one group of organic substances to another than does G_i (Berner, 1978a; 1980).

The concentrations G_i are also affected by non-biological processes. Readily decomposable materials (biopolymers) are broken down to small organic molecules which are not only converted by micro-organisms to CO_2, CH_4, and the like, but also can combine with one another abiogenically to form geopolymers. The geopolymers, which are less easily metabolized may, when biopolymers are exhausted, be at least partly decomposed. Thus, breakdown of the original biopolymers does not proceed entirely to the formation of simple inorganic molecules, and geopolymers (metabolized with difficulty) may be formed simultaneously and broken down. This overall process is illustrated in Figure 4-3.

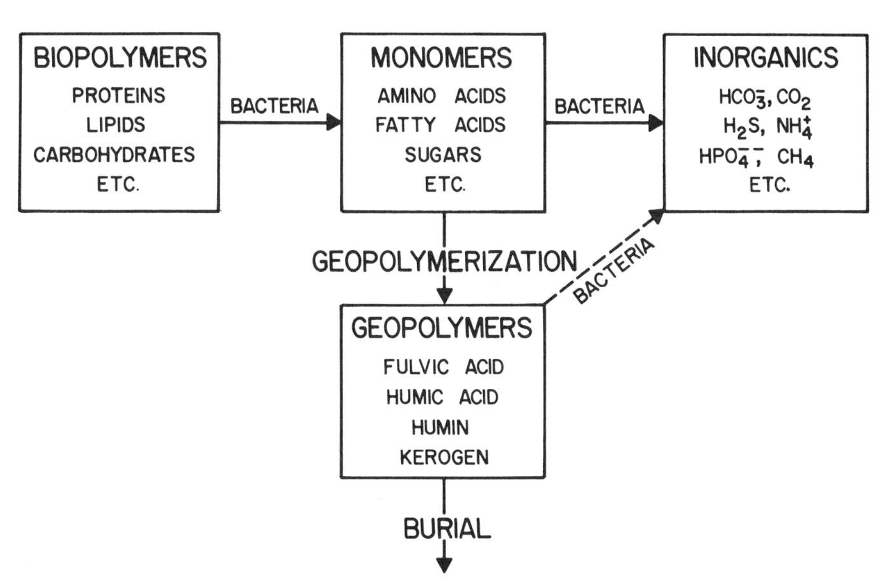

FIGURE 4-3. Idealized representation of the transformation of organic matter in sediments during early diagenesis. (See Welte, 1973.)

Unfortunately, little is known about the geopolymerization process. Our assumption in this book is that it is very rapid and is essentially completed near the sediment-water interface. Thus, we assume that further burial results in only bacterial decomposition, and that the geopolymers represent rather abundant (high G_i) but unreactive (low k_i) fraction(s) of the total decomposable organic matter (G_T). While this is a very rough assumption, it leads to useful results.

If one focuses on a single bacterial process, where there is unappreciable depletion of the most reactive organic substances present, we can treat the process in terms of only one group of compounds. In other words, for the 3-group equation above, if $k_1 \gg k_2 \approx k_3$ and G_1 does not become much less than G_2 or G_3, then we can assume that:

$$G_2 = G_3 = \text{constant},$$

and

$$\frac{dG_T}{dt} = \frac{dG_1}{dt} = -k_1 G_1. \tag{4-73}$$

This "one-G" type model has been extensively used by the writer to describe organic matter decomposition accompanying bacterial sulfate reduction (Berner, 1964; 1974), but more recent work (Jorgensen, 1978; Berner, 1980) suggests that in many sediments a "multi-G" model, such as that represented by equation (4-71), is a better representation. More work on this problem is sorely needed.

In addition to organic compounds, many inorganic species serve as reactants in microbial reactions. This is especially true when the inorganic species provide oxygen for bacterial oxidation of organic compounds. Common examples, already alluded to above, are dissolved O_2, SO_4^{--}, and NO_3^-. Rates of change of these species, due to microbial reduction, can also be expressed in terms of the Michaelis-Menten equation. For example, for dissolved O_2:

$$\frac{dC_{O_2}}{dt} = -\frac{R_{max} C_{O_2}}{K'_m + C_{O_2}} \tag{4-74}$$

Here, because we are dealing with a dissolved species, the use of the Michaelis-Menten equation is more justified than in the case of insoluble organic matter discussed above. However, rates of consumption of inorganic oxidants are also dependent on the concentrations, G, of organic substances reacting with the oxidants, and, in general, we must deal with two-reactant processes, such as that represented by equation (4-66). In sediments where organic matter (expressed as mass of carbon per unit volume of pore water) is present at much greater concentrations than the inorganic oxidants, two-reactant equations might possibly be represented by simple Michaelis-Menten type equations (e.g., see reasoning leading to equation 4-67). Thus, in the case of O_2 consumption via first order one-G type kinetics with $G \gg C_{O_2}$, R_{max} in equation (4-74) is given by:

$$R_{max} = k'G = \text{constant}, \tag{4-75}$$

where $k' = k\bar{\rho}_s(1 - \phi)/\phi$. Much more work is needed on the testing and application of equations like (4-74) to the various natural oxidants. In the case of sulfate it appears that rates of reduction are relatively independent of sulfate concentration and dominated instead by changes in G and k (Berner, 1979).

Rate expressions for microbial reactions need not rest on *a priori* models such as those described above. If a microbial reaction is the major chemical process affecting a metabolite in a sediment, it is possible to deduce the rate and concentration dependency of the reaction in terms of sediment data alone. For example, consider dissolved inorganic species in a sediment undergoing molecular diffusion, depositional advection, and bacterial removal. Also, assume simple linear adsorption, no bioturbation or compaction, and that diagenesis is at steady state. Under these conditions:

$$\left(\frac{D_s}{1 + K}\right)\frac{\partial^2 C}{\partial x^2} - \omega\frac{\partial C}{\partial x} - \frac{R}{1 + K} = 0, \tag{4-76}$$

so that solving for R:

$$R = D_s\frac{\partial^2 C}{\partial x^2} - (1 + K)\omega\frac{\partial C}{\partial x}. \tag{4-77}$$

Thus, if the concentration vs depth profile is measured, and values of D_s, K, and ω are estimated or determined directly, it is possible to calculate R at each depth by graphical differentiation. This method has been shown to be useful for sulfate profiles (below the zone of bioturbation) by Goldhaber et al. (1977). (See also Chapter 6.) In fact, if the profiles are exponential and approach a constant concentration with depth, one can show directly (Berner, 1979) that organic matter is being decomposed via first order one-G type kinetics.

5

Diagenetic Chemical Processes II: Precipitation, Dissolution, and Authigenic Processes

In this chapter, chemical processes which bring about the formation and destruction of minerals in a sediment during early diagenesis are discussed. Again, the approach is theoretical, and heavy emphasis is placed on the kinetics of crystal growth and dissolution. In addition, separate discussion of the means of emplacement and replacement of new minerals, here termed authigenic processes, is presented in terms of diagenetic modeling. Finally, there is a brief qualitative account of some of the causative chemical factors which give rise to authigenic processes (specifically cementation and concretion formation).

Precipitation

Under this heading we will discuss the fundamental principles which govern the crystallization of solids from aqueous solution. Although the specific goal is to improve our understanding of authigenic (diagenetic) mineral formation, the equations developed here apply to precipitation in general. Much of the following discussion is patterned after that of Nielsen (1964), and the reader is referred to this book as well as that by Ohara and Reid (1973) for further details.

Energetics of Precipitation

The overall process of mineral precipitation can be divided into two stages, nucleation and crystal growth. Distinction between the two can readily be made once one has a proper understanding of the energy changes which accompany crystallization. Consider an atom (ion, molecule) at the surface of a crystal. It has a greater energy than a similar atom within the body of the crystal because it is bound to adjacent atoms only on one side. Thus, it takes excess energy to move an atom from the interior to the surface of a crystal or, in other words, it takes excess energy to create new surface area. One mole of fine crystals of a given substance has more energy than one mole of coarse crystals. This excess energy, when combined with surface entropy effects, is known as interfacial free energy. The term "interfacial free energy" is used rather than the more common "surface free energy" because the energy of atoms depends on the nature of the material which is contacted across the interface. In other words, the interfacial energy of a crystal against an aqueous solution is different from its interfacial energy against another crystal or against air.

For crystals larger than about 2μm the interfacial free energy is very small relative to the bulk free energy, and it can usually be ignored. This means that in the usual thermodynamic calculations involving only bulk free energies, such as those presented in the previous chapter, it is implicitly assumed that all solid reactants and products are reasonably coarse-grained. On the other hand, when one considers the initial formation of crystals by precipitation, the size is so small that interfacial free energy becomes dominant. This can be seen from the following calculation. For the precipitation of a single crystal from solution we have:

$$\Delta G_n = \Delta G_{bulk} + \Delta G_{interf}, \tag{5-1}$$

where ΔG_n = free energy of formation of the crystal, and the subscripts *bulk* and *interf* refer respectively to the bulk and interfacial contributions to the free energy. If Ω refers to the ratio of the initial ion activity product of the supersaturated solution to the ion activity product at equilibrium (see equation 4-22), then:

$$\Delta G_{bulk} = -nk_B T \ln \Omega, \tag{5-2}$$

where n = number of atoms or ions precipitated to form the crystal;
k_B = Boltzmann constant;
T = absolute temperature in °K.

Also, we may define the important parameter σ:

$$\sigma = dG_{interf}/dA, \tag{5-3}$$

where σ = specific interfacial free energy between crystal and aqueous solution;
A = surface area of the crystal.

Some estimated values of σ for solids are shown in Table 5-1. If σ is not a function of A (a necessary assumption for all that follows), then one obtains:

$$\Delta G_{interf} = \sigma A, \tag{5-4}$$

TABLE 5-1

Interfacial free energies of some solids with respect to water.

Substance	σ (*erg cm*$^{-2}$)	*Reference*
$CaCO_3$ (calcite)	80	Berner and Morse (1974)
$SrCO_3$	92	Stumm and Morgan (1970)
$BaCO_3$	115	Nielsen and Söhnel (1971)
$PbCO_3$	125	" " " "
$BaSO_4$	135	" " " "
$SrSO_4$	85	" " " "
$CaSO_4 \cdot 2H_2O$	76	" " " "
KCl	30	" " " "
SiO_2 (amorph)	46	Stumm and Morgan (1970)
Glycine (crystals)	29	" " " "
ZnO	770	" " " "
CuO	690	" " " "
$Cu(OH)_2$	410	" " " "
Fe_2O_3	1200	Langmuir (1971)
$HFeO_2$	1600	" "

Substituting (5-2) and (5-4) in (5-1):

$$\Delta G_n = -nk_BT \ln \Omega + \sigma A. \tag{5-5}$$

Now we have:

$$A = bV^{2/3}, \tag{5-6}$$

$$V = nv_n, \tag{5-7}$$

where V = volume of the crystal;
b = geometrical constant (shape factor);
v_n = volume of an atom or ion in the crystal.

Substituting (5-6), and (5-7) in (5-5):

$$\Delta G_n = -nk_BT \ln \Omega + \sigma b v_n^{2/3} n^{2/3}. \tag{5-8}$$

This equation expresses the free energy of formation of a single crystal from supersaturated solution as a function of the degree of saturation Ω and the number of atom or ions in the crystal n. It is plotted at different arbitrary values of Ω in Figure 5-1.

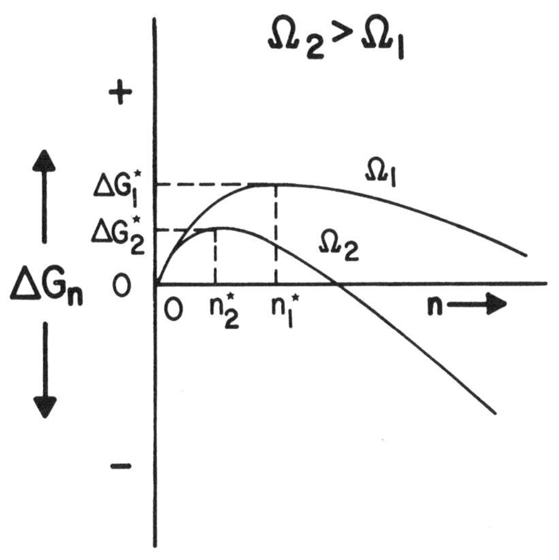

FIGURE 5-1. Plot of free energy for the formation of a single crystal ΔG_n as a function of the number of atoms (ions, molecules) in the crystal n. Note the free energy maximum necessary for nucleation. Also, at higher Ω values the free energy of nucleation ΔG_n^* and the size of the critical nucleus n^* are both smaller. (Adapted from Nielsen, 1964.)

NUCLEATION

The most striking fact exhibited by Figure 5-1 is that the formation of a crystal by precipitation from solution initially requires an increase in free energy. This is because the surface free energy term in equation (5-8), with its $\frac{2}{3}$ dependence on n, dominates at low values of n. The process during which the maximum in free energy is attained is known as *nucleation*, and involves the growth of tiny sub-microscopic *embryos* (which are unstable relative to re-solution). The energy necessary to reach the maximum and form the critical nucleus ($n = n^*$) is known as the free energy of nucleation ΔG_n^*. Once the critical nucleus is formed, further increase in n is accompanied by a decrease in free energy (due to dominance of the first term in equation 5-8), and this spontaneous process is known as *crystal growth* with the precipitating bodies now denoted as *crystals*. Increase of Ω, as shown in Figure 5-1, causes a decrease in both the free energy of nucleation and the size of the critical nucleus (as reflected by n^*).

The point of maximum free energy in Figure 5-1 is a state of chemical equilibrium (albeit unstable equilibrium) where the solution with a given value of Ω can be considered as being just saturated with respect to crystals of the size of the critical nucleus. In other words, fine crystals have excess solubility due to excess surface free energy and this excess solubility can be calculated directly from equation (5-8). The same calculation can, alternatively, be viewed as determination of the size of the critical nucleus from a known value of Ω. Calculation proceeds as follows. At the maximum:

$$\frac{d(\Delta G_n)}{dn} = 0, \tag{5-9}$$

and:

$$-k_B T \ln \Omega + (\tfrac{2}{3})\sigma b v_n^{2/3} n^{-1/3} = 0. \tag{5-10}$$

Introducing the nominal crystal radius, r_c:

$$r_c = \frac{b'V}{A}, \tag{5-11}$$

we have from equations (5-6), (5-7) and (5-11):

$$n^{-1/3} = \frac{b' v_n^{1/3}}{b r_c}, \tag{5-12}$$

so that substituting (5-12) in (5-10) and rearranging we obtain the size, r_c^*, of the critical nucleus:

$$r_c^* = \frac{(\tfrac{2}{3})\sigma b' v_n}{k_B T \ln \Omega}, \tag{5-13}$$

or alternatively the "solubility" of a crystal of radius r_c:

$$\ln \frac{K(r_c)}{K(\infty)} = \ln \Omega = \frac{2\sigma b' v}{3RT}\left(\frac{1}{r_c}\right), \tag{5-14}$$

where v_n and k_B are replaced by v (the molar volume) and R (the gas constant), respectively. In equation (5-14), Ω is replaced by the ratio of the equilibrium ion activity product for radius r_c to the equilibrium ion activity product for "infinitely large" ($r_c > 2\mu$m) crystals. In either case equations (5-13) and (5-14) are referred to as the Kelvin Equation.

Nucleation may occur either homogeneously or heterogeneously. In homogeneous nucleation the critical nucleus is formed purely by chance collisions of embryos with atoms or ions in solution, and the whole process can be considered as resulting from entropy fluctuations. In heterogeneous nucleation, the critical nuclei form on seed crystals which provide a considerable portion of the excess energy needed for nucleation. The surface of the seeding material can be thought of as a template of similar atomic spacing which promotes further precipitation. In fact, if the precipitate and seed consist of the same mineral, the nucleation energy is minimal and the overall process can be considered essentially as pure crystal growth. In sediments, because of a very high density of potential seeds, it is likely that nucleation is almost always heterogeneous.

The rate of homogeneous nucleation can be calculated (Nielsen, 1964) according to the expression:

$$N = \bar{\nu} \exp\left[\frac{-4b^3\sigma^3 v_n^2}{27(k_B T)^3(\ln \Omega)^2}\right], \tag{5-15}$$

where N = number of nuclei formed per unit volume per unit time;

$\bar{\nu}$ = a complex expression known as the "pre-exponential factor" which for our purposes can be assumed equal to 10^{31} nuclei per cm^3 per sec. (for a fuller discussion consult Nielsen, 1964).

In this expression there is a strong dependence of nucleation rate on degree of supersaturation Ω, and extremely strong dependence upon specific interfacial free energy, temperature, and shape factor. For a given supersaturation, temperature, and shape, equation (5-15) shows how reducing the value of σ, via heterogeneous nucleation, can greatly accelerate nucleation. (This argument is only qualitative, however, since expressions for the rate of heteronucleation are more complicated than is represented by equation 5-15.)

The size of crystals formed in a sediment depends upon the relative rates of nucleation and crystal growth. At high rates of nucleation (due to high supersaturation) excess dissolved material may appear almost entirely as critical nuclei which have not had a chance to grow. For most sedimentary substances, the size of the critical nucleus is on the order of a few tens to a few hundreds of angstroms. Thus, rapid precipitation can result in the formation, via nucleation, of very fine-grained precipitates which are amorphous or poorly diffracting toward x-rays. An extreme example is the x-ray amorphous "gel" of ferric hydroxide which is formed by rapid nucleation upon mixing solutions of dissolved ferric and hydroxide ions in the laboratory.

If nucleation is sufficiently slow, crystal growth becomes the dominant process of precipitation. In this case, for a given amount of excess dissolved material, there are fewer nuclei and, therefore, fewer and larger crystals formed. For a given substance, this situation is favored by low degrees of supersaturation and/or by uninhibited crystal growth (see below).

CRYSTAL GROWTH

Crystal growth involves (1) the transport of ions (atoms, molecules) to the surface of a crystal, (2) various surface reactions (adsorption, surface nucleation, surface diffusion, dehydration, ion exchange, etc.) that result

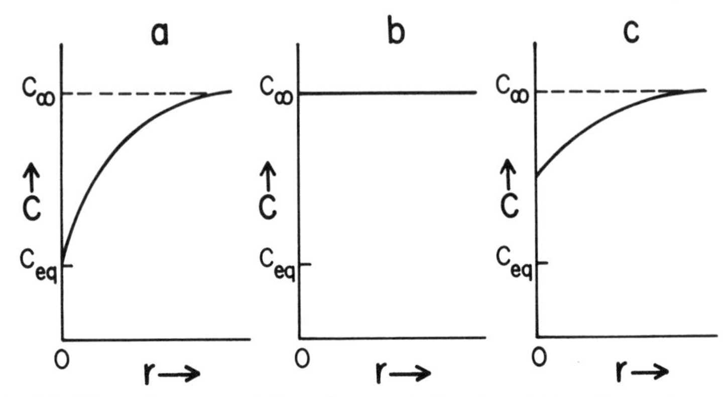

FIGURE 5-2. Schematic representation of concentration in solution C as a function of radial distance r, from the surface of a growing crystal. C_{eq} = saturation concentration, C_{∞} = concentration out in solution.
(a) Transport control.
(b) Surface-reaction control.
(c) Mixed transport and surface-reaction control.

in incorporation of the ions into the crystal lattice, and (3) removal of products of the reaction (if any) from the crystal. Ignoring product removal, the rate of growth may be limited either by transport, by surface chemical reaction, or by a combination of both processes. A comparison of the three rate-controlling mechanisms is shown in Figure 5-2.

In pure transport-controlled growth (Figure 5-2a) ions are attached so rapidly to the surface of the crystal that the concentration in solution immediately adjacent to the crystal is lowered to essentially the equilibrium or saturation level. Growth is then limited by the rate at which ions can migrate to the surface via diffusion and advection. The rate of growth, consequently, depends upon hydrodynamic conditions in solution, with faster growth resulting from increased flow velocities or increased stirring.

By contrast, pure surface-reaction controlled growth (Figure 5-2b) results when attachment via surface reactions is so slow that concentrations adjacent to the surface build up to values essentially the same as in the surrounding solution. Rate of growth is limited by surface reactions and is not affected by increased flow velocities in the external solution.

Situations can arise where surface reactions are sufficiently fast that ion depletion occurs adjacent to the crystal surface but transport prevents the concentration from being lowered to the saturation value. As a result, an intermediate concentration at the surface results. In this case we are dealing with mixed transport-surface reaction controlled kinetics. This situation is depicted in Figure 5-2c.

At water flow rates found in sediments, crystal growth via transport-controlled kinetics for a given substance should always be faster than that via surface-reaction control.* Also, rates of surface attachment are a strong function of the degree of supersaturation. As a result, as supersaturation is relieved by precipitation, the concomitant slowing of attachment rate may cause a change in rate-controlling mechanism from transport control at high supersaturation to surface reaction control at low supersaturation. This situation of changing rate control can be common during the approach to equilibrium (Nielsen, 1964). Also, the slowest rate at which a crystal may grow via transport control is where flow velocities are zero, i.e., where the water is stagnant. In this case all transport of material to the surface of a growing crystal must occur via molecular diffusion. (For fine-grained sediments below the zone of bioturbation this may be the appropriate situation.) Growth via molecular diffusion constitutes a limiting situation. Slower rates must be due to surface-reaction control and faster rates to transport control via flow of pore water. Calculation of the rate of precipitation via molecular diffusion, and comparison with measured rates, thus makes possible deduction of the rate-controlling process.

TRANSPORT-CONTROLLED GROWTH

The rate of growth of a crystal via molecular diffusion can be calculated theoretically (Frank, 1950; Nielsen, 1961) if the crystal can be approximated by a sphere. The reasoning is as follows. Assume that a sphere of radius r_c grows from solution by radially symmetric inward diffusion. Fick's Second Law of diffusion for spherical symmetry (equation 3-41) with constant diffusion coefficient is:

$$\frac{\partial C}{\partial t} = D_s \left[\frac{\partial^2 C}{\partial r^2} + \frac{2}{r} \frac{\partial C}{\partial r} \right], \tag{5-16}$$

where r = radial coordinate;
t = time;
D_s = molecular diffusion coefficient in a sediment pore water.

Solution of this equation for constant r_c and the boundary and initial conditions:

$$C(r,0) = C_\infty,$$
$$C(r_c,t) = C_{eq},$$
$$C(\infty,t) = C_\infty,$$

* At extremely high flow rates, as produced in the laboratory but not in nature, surface attachment can become rate limiting in otherwise transport-controlled growth—see Berner (1978b).

yields:

$$C = C_{eq} + (C_\infty - C_{eq})\left\{1 - \left(\frac{r_c}{r}\right)\left(1 - erf\left[\frac{(r - r_c)}{2\sqrt{Dt}}\right]\right)\right\}. \quad (5\text{-}17)$$

As $[(r - r_c)/2\sqrt{(Dt)}] \to 0$, equation (5-17) reduces to the stationary state expression:

$$C = C_{eq} + (C_\infty - C_{eq})(1 - r_c/r). \quad (5\text{-}18)$$

It can be shown (Nielsen, 1961) that for almost all minerals the stationary state is set up much faster than the sphere grows, so that the assumptions leading up to equation (5-18) are justified. In this case, we can combine (5-18) with Fick's First Law (for spherical symmetry) in a sediment:

$$J = -\phi D_s \frac{\partial C}{\partial r},$$

$$J_{r=r_c} = -\phi D_s \left(\frac{\partial C}{\partial r}\right)_{r=r_c} = \frac{-\phi D_s(C_\infty - C_{eq})}{r_c}, \quad (5\text{-}19)$$

where J = diffusive flux;

ϕ = sediment porosity.

Now the sphere grows by addition of the material diffusing to its surface such that:

$$\frac{dM}{dt} = -4\pi r_c^2 J_{r=r_c}, \quad (5\text{-}20)$$

$$\frac{dr_c}{dt} = \frac{v}{4\pi r_c^2 \phi_c}\frac{dM}{dt}, \quad (5\text{-}21)$$

where M = mass of the growing sphere;

v = molar volume of precipitating material;

ϕ_c = volume fraction of sediment occupied by precipitating material (for cementation $\phi_c = \phi$).

If $\phi_c = \phi$, then from (5-19), (5-20), and (5-21):

$$\frac{dr_c}{dt} = \frac{v D_s(C_\infty - C_{eq})}{r_c}. \quad (5\text{-}22)$$

Equation (5-22) is the proper expression for the diffusion-controlled growth of a spherical shaped crystal. In terms of a large number of crystals it can be recast, using the relation:

$$\frac{dC}{dt} = -n'\frac{dM}{dt}, \tag{5-23}$$

where n' = number of crystals per unit volume of pore water,

in the form:

$$\frac{dC}{dt} = \frac{-\bar{A}\phi D_s(C_\infty - C_{eq})}{r_c}, \tag{5-24}$$

where $\bar{A} = 4\pi r_c^2 n'$ (the surface area of growing crystals per unit volume of pore water).

This is the expression that is used to represent diffusion-controlled crystal growth (or dissolution—see below) in diagenetic equations. It is strictly true only if the growing crystals are about the same size and are separated by at least five diameters on the average (Nielsen, 1964).

If water flow in a sediment is appreciable, equations (5-22) and (5-24) need to be modified to account for the additional flux of material to the surface of the crystal via advection. Nielsen (1961) has derived a theoretical expression for the transport-controlled growth of a sphere from flowing water, with the assumption of uniform laminar flow. His resulting equation can be expressed as:

$$\frac{dr_c}{dt} = \frac{\upsilon D_s(C_\infty - C_{eq})}{r_c}\left(1 + \frac{r_c U}{D_s}\right)^{0.285}, \tag{5-25}$$

where U = (uniform) flow velocity. For a large number of spherical crystals:

$$\frac{dC}{dt} = -\frac{\bar{A}\phi D_s(C_\infty - C_{eq})}{r_c}\left(1 + \frac{r_c U}{D_s}\right)^{0.285}. \tag{5-26}$$

An alternative expression to (5-26), for transport-controlled growth from flowing water, is the empirical expression, sometimes referred to as the Frössling equation (Ohara and Reid, 1973):

$$\frac{dC}{dt} = -\frac{\bar{A}\phi D_s(C_\infty - C_{eq})}{r_c}\left[1 + 0.42\frac{(r_c U)^{1/2}}{(\nu D_s^2)^{1/6}}\right], \tag{5-27}$$

where ν = kinematic viscosity.

Whether or not equations (5-26) or (5-27), which are both based on the flow of water past a single sphere, can be applied directly to sediments is debatable, but at the slow flow rate and fine grain size found in sediments, the flow correction factors (terms on right-hand side in brackets) are close to one anyway. For example, for 100 μm crystals subject to a (relatively high) water flow rate of 30 meters per year, at typical values of v and D_s, the values of the correction terms in brackets are 1.09 and 1.06 for the Nielsen and Frössling expressions respectively.

Transport-controlled crystal growth in vigorously stirred solution is sometimes described in terms of a uniform, stagnant boundary layer around each crystal, which is referred to as the Nernst Layer. This concept has been largely discredited by recent work (e.g., Levich, 1962) where fluid motion has been demonstrated well within the supposedly stagnant layer. Moreover, it is inappropriate as a geometrical description of crystal growth within sediments where vigorous stirring does not occur. It will not be used here.

In the above expressions the symbol C has been used simply to represent the concentration of a dissolved species which can precipitate to form new crystals. Since almost all crystals of geological interest consist of more than one kind of ion, atom, or molecule, we need to redefine C for minerals more carefully. We can arbitrarily choose one of the ions of the crystal to be represented by C, but then the value of C_{eq} depends on the concentrations of other ions at the surface of the crystal. In other words, for a binary salt AB for example:

$$C_{eq_A} C_{eq_B} = K_c, \tag{5-28}$$

where K_c = concentration solubility product (see equation 4-19).

Determination of C_{eq_A} requires knowledge of C_{eq_B} at the surface of the crystal. Methods of calculating surface equilibrium concentrations are given by Nielsen (1964) and Berner and Morse (1974). A simple example of the kind of calculation involved is for the case of a 1:1 salt AB where neither ion undergoes dissociation or association to form new ions. In this case, for mass and charge balance:

$$J_{A^+_{r=r_c}} = J_{B^-_{r=r_c}}. \tag{5-29}$$

Therefore, from equation (5-19):

$$D_{s_{A^+}}(C_{\infty A^+} - C_{eq_{A^+}}) = D_{s_{B^-}}(C_{\infty B^-} - C_{eq_{B^-}}). \tag{5-30}$$

If $D_{s_{A^+}} \approx D_{s_{B^-}}$:

$$C_{eq_{B^-}} \approx C_{\infty B^-} - C_{\infty A^+} + C_{eq_{A^+}}. \tag{5-31}$$

Also:

$$C_{eq_{A^+}} = K_c / C_{eq_{B^-}}. \tag{5-32}$$

Thus, in the expression for transport-controlled growth, upon substituting (5-31) and (5-32) we obtain:

$$(C_{\infty A^+} - C_{eq_{A^+}}) = \left(\frac{C_{\infty A^+} + C_{\infty B^-}}{2}\right) - \frac{[(C_{\infty B^-} - C_{\infty A^+})^2 + 4K_c]^{1/2}}{2}. \tag{5-33}$$

Equation (5-33) can be considerably simplified for two situations. If $|C_{\infty B^-} - C_{\infty A^+}| \ll K_c^{1/2}$:

$$(C_{\infty A^+} - C_{eq_{A^+}}) \approx (ICP)^{1/2} - K_c^{1/2}, \tag{5-34}$$

where $ICP = C_{\infty A^+} C_{\infty B^-}$.

If $C_{\infty B^-} \gg C_{\infty A^+}$:

$$(C_{\infty A^+} - C_{eq_{A^+}}) \approx C_{\infty A^+}. \tag{5-35}$$

SURFACE-REACTION CONTROLLED GROWTH

If the rate of crystal growth is limited by surface reactions, we are no longer concerned with the transport properties of the surrounding solution. Instead the problem reduces to one of trying to formulate the slowest or rate-controlling step at the surface. This is not a simple problem and, as a result, many theories of surface-reaction controlled growth have been developed (see Ohara and Reid, 1973, for a detailed discussion). Here we will assume, as in most theories, that rates are controlled by surface nucleation and/or surface diffusion. Two types of growth are distinguished: surface-nucleation controlled growth and dislocation-controlled growth.

In surface-nucleation controlled growth, the rate-limiting step is formation of a flat, two-dimensional crystal (critical nucleus) one ion (atom, molecule) thick, on an otherwise atomically smooth crystal surface (Figure 5-3). The new crystal provides atomic sized *steps* which greatly promote attachment of new ions. This is because, for statistical reasons, there are a certain number of *kinks* or favored growth sites along the step (see Figure 5-3). At a kink more bonds are available for the attachment of the ions than are available at straight sections of the step or, especially, on atomically flat portions of the crystal surface. Because of preferred attachment at kinks (which produces new kinks), the whole process of crystal growth can be visualized as the migration of kinks and steps. In order to provide steps and kinks, according to the surface-nucleation model, a new nucleus must be formed. The rate by which the surface nucleus forms can be calculated from standard nucleation theory. As

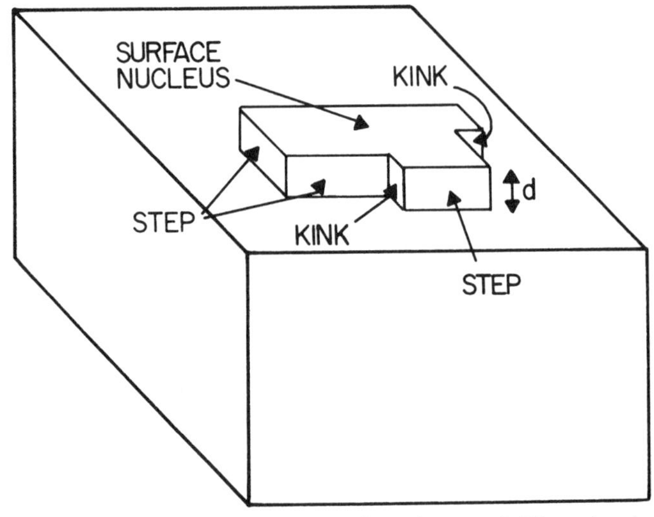

FIGURE 5-3. Idealized representation of the surface of a crystal. Dimension d represents one atom, molecule, unit cell, etc. On the flat crystal surface a flat ("two-dimensional") surface nucleus is present which exhibits monoatomic steps and kinks.

derived by Nielsen (1964), one simplified formulation is:

$$\ln N' = \ln\left(\frac{D_s}{v_n^{4/3}}\right) - \frac{\beta' \sigma^2 v_n^{4/3}}{(k_B T)^2 \ln \Omega}, \tag{5-36}$$

where N' = surface-nucleation rate in nuclei per unit area per unit time;

β' = two-dimensional shape factor (for a circular disc it is equal to π);

v_n = atomic volume (molar volume/6×10^{23}).

According to the *mononuclear* theory (Nielsen, 1964; Ohara and Reid, 1973), the rate of growth is limited solely by surface nucleation. In other words, once a surface nucleus has formed, the rate of spreading of its steps by ion attachment is so fast that an entire monoatomic crystal layer is formed before another surface nucleus is created. Each atomic layer on the crystal surface is thus formed from only one surface nucleus. Under these conditions (assuming all crystal faces grow at the same rate) the rate of growth is:

$$\frac{dr_c}{dt} = N' A_c d, \tag{5-37}$$

where A_c = surface area of the whole crystal;
d = atomic diameter (5×10^{-8}cm).

If the crystal maintains its shape during growth:

$$A_c = b''r_c^2, \tag{5-38}$$

where b'' is a geometric constant. Combining (5-37) and (5-38), the rate law for mononuclear growth is obtained:

$$\frac{dr_c}{dt} = k_m r_c^2, \tag{5-39}$$

where $k_m = N'b''d$ = mononuclear rate constant.

Mononuclear growth is rather unlikely and has not been documented for any given substance. A more probable situation is that of polynuclear growth (Nielsen, 1964; called "birth and spread" model by Ohara and Reid, 1973). In polynuclear growth, nucleation is still two-dimensional following equations such as (5-36), but more than one nucleus forms before an atomic layer can spread to cover an entire crystal face. Thus, nuclei can form on partially completed atomic layers or on flat sections which have not yet been reached by step spreading from another nucleus. Polynuclear growth results from assuming that spreading of steps is not infinitely fast, but rather is limited by surface diffusion. Several different formulations of the polynuclear model have been made (Nielsen, 1964; Ohara and Reid, 1973), but all have in common the same zeroth order rate dependence on r_c. In other words:

$$\frac{dr_c}{dt} = k_p, \tag{5-40}$$

where k_p = polynuclear rate constant whose value depends on the particular model adopted.

A major problem with both surface nucleation theories is that the rate dependence upon the degree of supersaturation is predicted to be much, much, higher than has been found in laboratory experiments. The high dependence on Ω is predictable from the rate of nucleation itself, which is strongly dependent on Ω as shown by equation (5-36). (For typical values, $D_s = 3 \times 10^{-6}$ cm^2 sec^{-1}, $v_n = 6 \times 10^{-23}$ cm^3, $\beta' = 3.14$, $\sigma = 100$ erg cm^{-2}, $T = 298$°K, and $\Omega = 2$, from equation (5-36) we obtain $\partial(\ln N')/\partial(\ln \Omega) = 92$.) Thus, most surface-reaction controlled crystal growth must take place via some other mechanism.

A reasonable mechanism has been postulated to be growth at intersections of screw dislocations with the surface of the crystal (dislocation-controlled growth). Screw dislocations are common in all crystalline substances and, since they represent atomic-sized displacements along the axis of a row of atoms, their "outcrops" with the crystal surface must provide built-in atomic sized steps (see Figure 5-4). Thus, no nucleation of new steps is needed since the screw dislocation outcrops are always present. (This explains observations of crystal growth at low values of Ω (<1.01) where two-dimensional surface nucleation is virtually impossible because of insufficient supersaturation.) Surface crystals nucleated at an outcrop rotate in spiral-like fashion around the outcrops while maintaining a step for attachment at the front of the spiral (Figure 5-4). This spiral growth mechanism has been incorporated in the classical Burton, Cabrera, and Frank (BCF) Theory (Burton et al., 1951) which treats growth as being limited by surface diffusion into the spiral step. This results in an expression for dislocation-controlled growth which is rather complicated, but which can be simplified at low degrees of supersaturation ($\Omega < 1.01$) to:

$$\frac{dr_c}{dt} = k_D(C_\infty - C_{eq})^2, \tag{5-41}$$

where k_D = dislocation rate constant (a complex function of many variables).

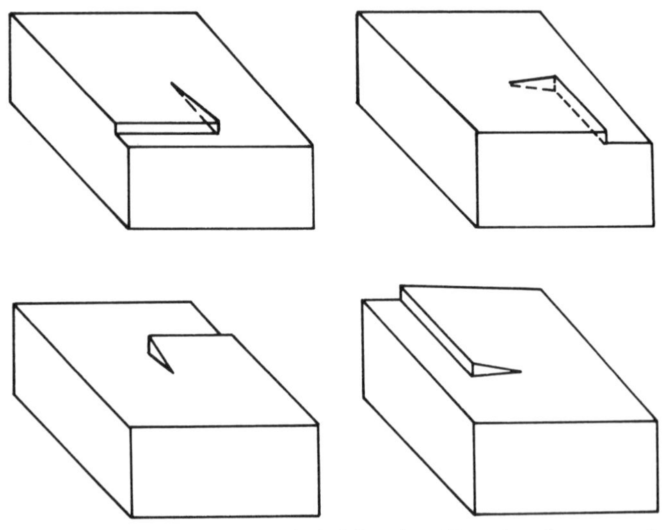

FIGURE 5-4. Step migration along a spiral dislocation. Note that the step at the growth front is self-perpetuating. (Modified from Nielsen, 1964.)

As supersaturation increases, the growth rate dependence varies between $(C_\infty - C_{eq})^2$ at very low supersaturation (as above), to $(C_\infty - C_{eq})$ at very high degrees of supersaturation.

The surface nucleation and dislocation models discussed here have been presented mainly as an aid to gaining a better understanding of major factors affecting crystal growth. It should not be construed that all or any actual cases of crystal growth in sediments are covered by these models. This is especially true where certain inhibitor ions (or poisons) are adsorbed on the crystals and, as a result, bring about a deceleration of growth by blocking active sites for attachment (e.g., kinks). In such cases different dependencies on $(C_\infty - C_{eq})$ may occur than are predicted by the BCF theory (Ohara and Reid, 1973). Since natural sediment pore waters contain a variety of dissolved species which could serve as inhibitors, it is likely that inhibition type surface-reaction controlled growth is important. Unfortunately, there is no widely accepted theory for growth of this type. Nevertheless, as a first approximation we can assume that surface-reaction controlled growth in the presence or absence of inhibitors follows a relation of the form:

$$\frac{dr_c}{dt} = k(C_\infty - C_{eq})^n, \qquad (5\text{-}42)$$

where $n > 1$. This relation has been shown to hold for the inhibited dissolution of $CaCO_3$ in seawater (Morse, 1978; Keir, 1979; and see below).

Dissolution

In many respects dissolution can be considered as the inverse of precipitation. Like precipitation, dissolution rate is controlled either by transport or by surface reactions. The equations presented for describing transport-controlled precipitation also apply to transport-controlled dissolution, the only difference being that $C_\infty < C_{eq}$ so that $dr_c/dt < 0$ in equation (5-22) and $dC/dt > 0$ in equation (5-24). Also, like precipitation, dissolution can be visualized as taking place by the migration of steps and kinks, and deceleration of dissolution may occur by the adsorption of inhibitor ions at the kinks.

Unlike precipitation, dissolution does not involve three-dimensional nucleation since the bodies to be dissolved are already present. However, nucleation on pre-existing atomically flat crystal surfaces of unit pits (one atom deep) can be visualized as a way of producing steps on the surface. If step spreading is fast and dissolution is limited by such unit pit nucleation, we have the dissolution analogue of surface-nucleation-controlled layer-by-layer growth. Also, like growth, pit nucleation is greatly aided by dislocation outcrops on the surface which originate in this case from both screw and edge dislocation. The dislocation outcrops provide the excess energy necessary for creating new pits and their accompanying steps. By continual deepening of the pits along their dislocations and lateral step spreading, microscopically visible etch pits may form. One aspect of dissolution which is unique to it is that corners and edges of crystals serve as built-in kinks and steps respectively.

Discernment of the rate-controlling mechanism during dissolution is easier than it is for precipitation. As in the case for precipitation, one can calculate rates for molecular diffusion control via equation (5-24) and compare them with measured rates. If the measured rates are slower, then dissolution is controlled by surface reactions. However, one can also deduce rate-controlling dissolution mechanisms through microscopic study of the surfaces of partially dissolved crystals. Distinct, crystallographically controlled etch pitting and ledge formation (a ledge is a microscopically visible pile-up of steps) are indicative of surface-reaction controlled dissolution, whereas a general rounding of the crystals indicates transport control. The general rounding occurs because of the ease of pit nucleation all over the surface or, in other words, etch pits are not confined to dislocations. Examples of the two types of morphology on the same mineral (calcite) are shown in Figure 5-5.

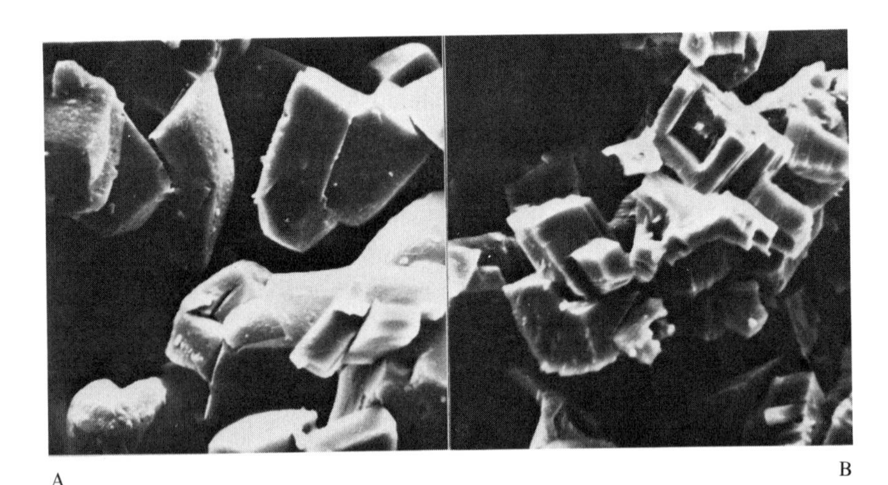

FIGURE 5-5. Electron photomicrographs of calcite which has undergone partial dissolution in seawater ($\times 5000$).
(A) Transport-controlled dissolution; pH = 3.9. Note general rounding.
(B) Surface-reaction controlled dissolution; pH = 6.0. Note angular, crystallographically controlled etch features.
(After Berner, 1978b and Berner and Morse, 1974.)

TABLE 5-2

Dissolution rate-controlling mechanism for various substances arranged in order of solubilities in pure water (mass of mineral which will dissolve to equilibrium). Data from Christoffersen et al. (1978) and a variety of references compiled by Berner (1978b).

Substance	*Solubility mole per liter*	*Dissolution rate control*
$Ca_5(PO_4)_3OH$	2×10^{-8}	Surface-reaction
$KAlSi_3O_8$	3×10^{-7}	Surface-reaction
$NaAlSi_3O_8$	6×10^{-7}	Surface-reaction
$BaSO_4$	1×10^{-5}	Surface-reaction
$AgCl$	1×10^{-5}	Transport
$SrCO_3$	3×10^{-5}	Surface-reaction
$CaCO_3$	6×10^{-5}	Surface-reaction
Ag_2CrO_4	1×10^{-4}	Surface-reaction
$PbSO_4$	1×10^{-4}	Mixed
$Ba(IO_3)_2$	8×10^{-4}	Transport
$SrSO_4$	9×10^{-4}	Surface-reaction
Opaline SiO_2	2×10^{-3}	Surface-reaction
$CaSO_4 \cdot 2H_2O$	5×10^{-3}	Transport
$Na_2SO_4 \cdot 10H_2O$	2×10^{-1}	Transport
$MgSO_4 \cdot 7H_2O$	3×10^{0}	Transport
$Na_2CO_3 \cdot 10H_2O$	3×10^{0}	Transport
KCl	4×10^{0}	Transport
$NaCl$	5×10^{0}	Transport
$MgCl_2 \cdot 6H_2O$	5×10^{0}	Transport

The author (Berner, 1978b) has shown that several common minerals (calcite, feldspar, opaline silica) all dissolve in natural waters via surface-reaction-controlled kinetics. A correlation between dissolution mechanism and solubility (see Table 5-2) was found which enables the prediction that the rate of dissolution of many other minerals should also be controlled by surface processes. Furthermore, the presence of potential inhibitors in natural waters, especially sediment pore waters, may well cause minerals to dissolve in nature by surface-reaction control even though laboratory experiments (in pure solutions) indicate a transport-controlled mechanism.

Retardation of dissolution by the adsorption of inhibitor ions on reactive sites such as kinks can lead to highly non-linear rate laws. Morse

(1978) and Keir (1979) have shown that calcite dissolution in seawater occurs according to the rate law:

$$\frac{dC}{dt} = k\bar{A}(C_{eq} - C)^n, \qquad (5\text{-}43)$$

where $n = 4$–6. This strong power dependence has been ascribed by Berner and Morse (1974) and Morse and Berner (1979) to the adsorption of phosphate ions at reaction sites on the calcite surface. Dissolution is brought about by the ability of retreating steps to penetrate, via curved embayments, between adsorbed phosphate ions which otherwise act to pin down the steps and, thereby, inhibit dissolution.

Authigenic (Mineral-Diagenetic) Processes

An authigenic mineral is one that forms within a sediment after burial, in other words, during diagenesis. It may crystallize in the original pore space of the sediment or it may fill pore space created by the dissolution of a pre-existing mineral. In the first case the process is known as *cementation* and in the latter as *replacement*. Both cementation and replacement may result in localized mineral segregations termed concretions. The material in the authigenic precipitate may be derived from the dissolution of minerals within the sediment (diagenetic redistribution) or be supplied from outside the sediment. Likewise, material dissolved during replacement may precipitate within the sediment or be entirely removed from it. Various textural criteria, on a microscopic scale, exist for the discrimination of cementation from replacement; for detailed information the reader is referred to standard works on the subject (e.g., Folk, 1974; Bathurst, 1971). Our concern here will not be with microscopic results but, rather, with an overall theoretical approach to the description of authigenic processes in general, and large-scale processes in particular.

Diagenetic Redistribution

Diagenetic redistribution involves dissolution of a given substance, its migration, and its reprecipitation within the same sediment or rock unit. Fisher (1978) in studying metamorphic rocks has presented a scheme whereby the rate-controlling steps in metamorphic or diagenetic redistribution can be deduced. The reasoning goes as follows (see Figure 5-6). If both dissolution and precipitation are transport controlled, sharp gradients in concentrations of dissolved species will exist between the surfaces of dissolving and precipitating particles. The sharp gradients are maintained because of the ease of attachment and detachment of ions to

TRANSPORT CONTROL

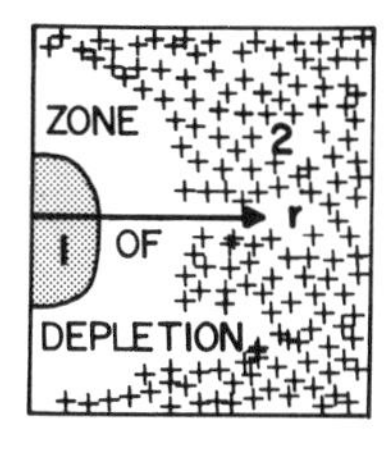

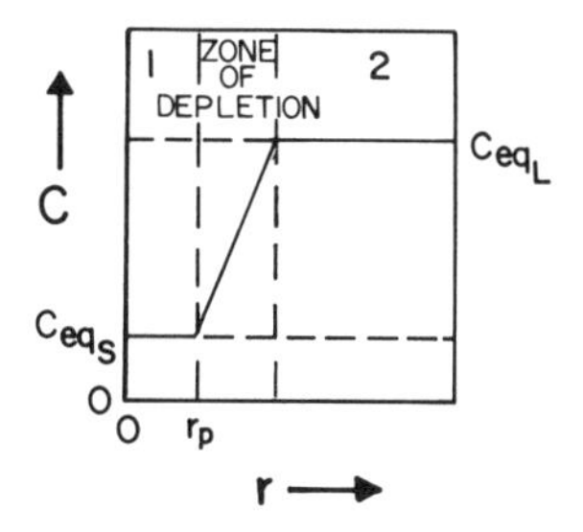

SURFACE REACTION CONTROL

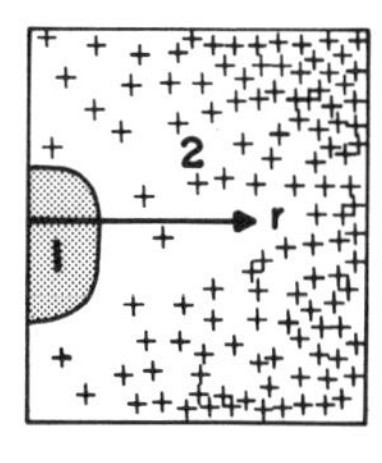

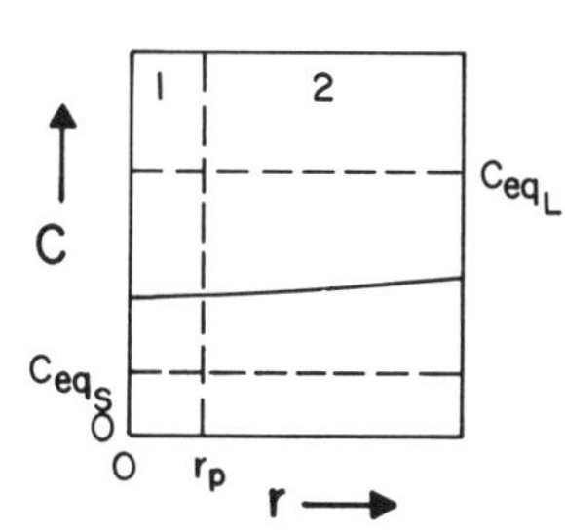

FIGURE 5-6. Schematic representation of diagenetic redistribution. Mineral 2, located in the sediment matrix, dissolves, and components of it re-precipitate to form mineral 1. The density of crosses in the region marked 2 are a measure of the concentration of mineral 2 in this region. Under transport control, sharp concentration gradients and a zone, around mineral 1, of total depletion of mineral 2, result. With surface-reaction control, concentration gradients are gentle and the zone of depletion around mineral 1 is diffuse. (Modified after Fisher, 1978.)

and from the crystal surfaces, and, as a result, equilibrium concentrations are present at both types of surface. Growth of each precipitating body is, therefore, at the expense of its nearest neighbors and as a result, a sharp, well-defined zone of total depletion is found around each growing body (Figure 5-6). The thickness of this zone depends on the concentration in the original sediment of the material involved in redistribution and its concentration in the precipitating body. For example, if a monomineralic segregation is forming in a sediment where its originally dispersed concentration is 1%, the volume of the zone of depletion, if transport control is operative, is one hundred times larger than the volume of the segretation. As the dissolving material increases in abundance, the zone of depletion becomes thinner until it practically disappears when the limit of 100% is reached. In this case we are, then, dealing with a special case of transport-controlled replacement.

By contrast with transport-controlled redistribution, if both dissolution and precipitation are surface-reaction controlled, detachment and attachment of ions are slow, and very low concentration gradients in solution result. In other words, considerable supersaturation exists at the surface of the precipitating bodies and considerable undersaturation at the surface of the dissolving particles. In this case dissolution becomes selective, and grains with excess energy, such as very small ones, dissolve preferentially even though they may be much further from a precipitating body than other less reactive grains. This gives rise to a diffuse zone of gradual depletion around each precipitating body (Figure 5-6) whose thickness cannot be predicted from simple mass balance arguments as was the case for transport-controlled redistribution.

The above discussion indicates that examination of the area around an accreting body, for example the spherical zone surrounding a concretion, can be used to decide whether or not the body was formed via local diagenetic redistribution and whether or not the process was controlled largely by transport. If transport control can be demonstrated, then it is possible to calculate the time it took to form the body. For example, consider the growth of a spherical concretion from dispersed, fine-grained material in the host rock (Figure 5-6 top). If the interstitial water beyond the zone of depletion is saturated with respect to the fine material, as it should be for transport control, and fluid flow is negligible compared to molecular diffusion, we have a simple case of the diffusion-controlled growth of a sphere, as described in the previous section. Because concentrations are fixed at the surface of the concretion and at the outer (spherical) surface of the zone of depletion, there is, at steady state, a simple linear concentration gradient radiating outward from the concretion. Thus, flux to the surface of the sphere is given by:

$$J_{r_p} = \frac{-\phi D_s}{L}(C_{eqL} - C_{eqS}), \tag{5-44}$$

where J_{r_p} = diffusion flux in mass per unit area per unit time at the surface of the sphere;

r_p = radius of the concretion;

C_{eqL} = equilibrium concentration at outer surface of the zone of depletion;

C_{eqS} = equilibrium concentration at surface of the sphere;

L = thickness of the zone of depletion.

(It is assumed here that steady state is rapidly re-established as the thickness of the zone of depletion and the radius of the concretion both increase, which is a reasonable assumption—see Nielsen, 1961 or Crank, 1975)

Now the concretion grows by addition of the surface flux according to:

$$\frac{dM_p}{dt} = -4\pi r_p^2 J_{r_p}, \tag{5-45}$$

where M_p = mass of the precipitating material.

Also:

$$dM_p = \frac{4\pi r_p^2}{v_p} F_p \, dr_p, \tag{5-46}$$

where v_p = molar volume of the precipitating material;
F_p = volume fraction in the concretion represented by the precipitating material.

Combining equations (5-44), (5-45), and (5-46):

$$\frac{dr_p}{dt} = \frac{\phi v_p D_s}{F_p L}(C_{eqL} - C_{eqS}). \tag{5-47}$$

At the same time that the concretion is growing, the thickness of the zone of depletion is also increasing:

$$\frac{dL}{dt} = \frac{dr_L}{dt} - \frac{dr_p}{dt}, \tag{5-48}$$

where r_L = radial distance to outer surface of the spherical depletion zone.

By reasoning analogous to that above for the concretion:

$$\frac{dM_d}{dt} = 4\pi r_L^2 J_L, \tag{5-49}$$

where M_d = mass of the dissolving material,
J_L = flux of dissolving material, across outer boundary of the depletion zone (J_L is equal to J_{r_p}),

and:

$$dM_d = \frac{-4\pi r_L^2}{v_d} F_d \, dr_L, \tag{5-50}$$

where F_d = volume fraction of dissolving material in bulk sediment,
v_d = molar volume of dissolving material,

combining (5-44), (5-49), and (5-50) and noting that $J_L = J_{r_p}$:

$$\frac{dr_L}{dt} = \frac{\phi \upsilon_d D_s}{F_d L}(C_{eqL} - C_{eqS}), \tag{5-51}$$

so that, from (5-47) and (5-48):

$$\frac{dL}{dt} = \frac{D_s(C_{eqL} - C_{eqS})}{L}\left(\frac{\phi}{F_d}\upsilon_d - \frac{\phi}{F_p}\upsilon_p\right). \tag{5-52}$$

Integrating equation (5-52) for the boundary conditions, $t = 0$, $L = 0$:

$$L = \left[2D_s(C_{eqL} - C_{eqS})\left(\frac{\phi}{F_d}\upsilon_d - \frac{\phi}{F_p}\upsilon_p\right)\right]^{1/2} t^{1/2}. \tag{5-53}$$

Now, we may also solve for the time required to grow the concretion to any radius r_p. Substituting (5-53) in (5-47):

$$\frac{dr_p}{dt} = 2^{-1/2} H D_s^{1/2}(C_{eqL} - C_{eqS})^{1/2} t^{-1/2}, \tag{5-54}$$

$$\text{where}\quad H = \frac{\phi \upsilon_p}{F_p}\left(\frac{\phi}{F_d}\upsilon_d - \frac{\phi}{F_p}\upsilon_p\right)^{-1/2}.$$

Integrating with the boundary conditions, $t = 0$, $r_p = 0$, and solving for t, we finally obtain the useful expression:

$$t = \frac{r_p^2}{2H^2 D_s(C_{eqL} - C_{eqS})}. \tag{5-55}$$

If the concretion forms by filling original pore space only (i.e., as a cement), $\phi = F_p$, and if also $\upsilon_d \approx \upsilon_p$, then H^2 simplifies to $\upsilon_p/(F_p/F_d - 1)$ so that equation (5-55) becomes:

$$t = \frac{r_p^2(F_p/F_d - 1)}{2\upsilon_p D_s(C_{eqL} - C_{eqS})}. \tag{5-56}$$

Thus, as the volume fraction of dissolving material becomes smaller and smaller, the time necessary for growth of the concretion becomes larger and larger. A numerical example of the use of equation (5-56) is shown in Figure 5-7 for the hypothetical case of calcite concretion growth from disseminated fine aragonite crystals. Note that transport-controlled diagenetic redistribution is relatively fast on a geological time scale.

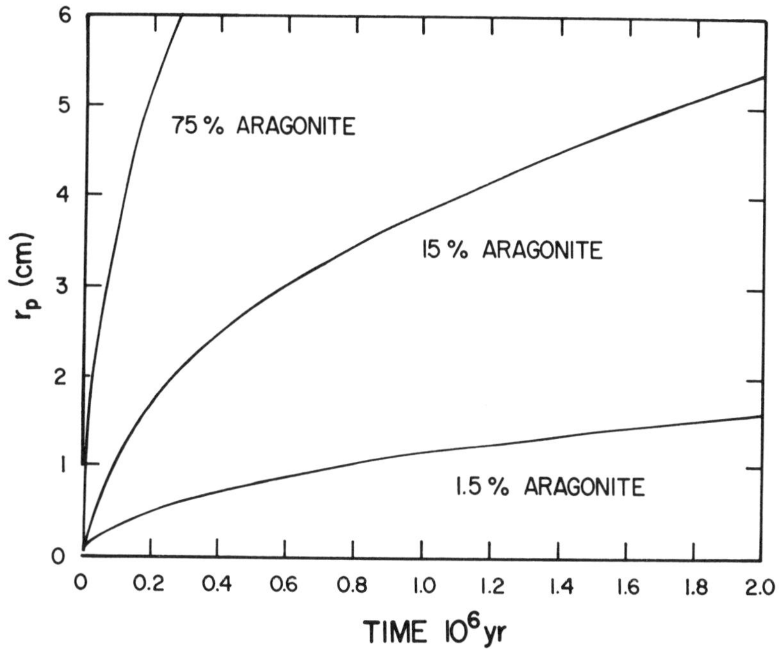

FIGURE 5-7. Plot of time necessary to grow spherical concretions of calcite by aragonite dissolution and transport (diffusion)-controlled diagenetic redistribution. The calcite concretions are assumed to consist of pore-filling cement. Values used for calculation: $(C_{eqL} - C_{eqS}) = 3.1 \times 10^{-5} M$ (for carbonate ion in seawater), $D_s = 2 \times 10^{-6}$ cm^2 sec^{-1}, $\phi = 0.60 = F_p$, $v_{\text{calcite}} \approx v_{\text{aragonite}} = 35$ cm^3 mol^{-1}. Aragonite percentages refer to volume % of total solids.

The reasoning used above can also be applied to the one-dimensional situation of the growth of a monomineralic layer in a sediment. Such a layer would be expected to form where vertical concentration gradients in the pore water are high. A good example is the formation of a layer of iron or manganese oxide at the oxygenated sediment-water interface of a sediment which is otherwise anoxic. Here dissolved Fe^{++} or Mn^{++} diffusing up from the underlying sediment are almost totally precipitated, by oxidation, to form Fe^{+3} or Mn^{+4} oxyhydroxides. If the Fe^{++} or Mn^{++} is derived via the anoxic dissolution of scattered fine particles of various iron and manganese minerals in the sediment, and if this dissolution is diffusion controlled, then we are dealing with transport-controlled diagenetic redistribution. (Precipitation at the sediment-water interface need not be diffusion controlled as long as it is sufficiently rapid, as it must be due to very high supersaturation in the presence of dissolved O_2.) Under

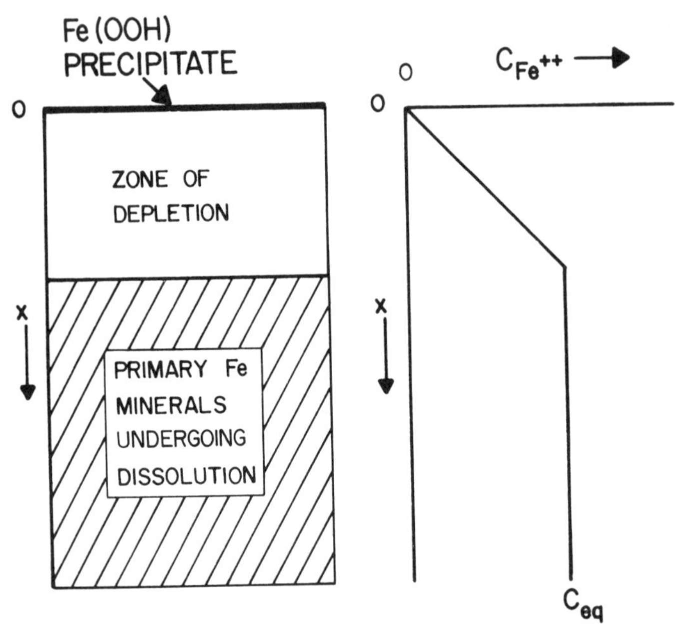

FIGURE 5-8. Schematic representation of formation of monomineralic layer of hydrous ferric oxide at the sediment-water interface as the result of transport- (diffusion) controlled diagenetic redistribution of iron (see text).

these conditions, a depletion zone of soluble Fe and Mn forms below the interface, and the concentration profile is linear across this zone (see Figure 5-8). Accordingly, the flux at $x = 0$ is given by equation (5-44):

$$J_0 = \frac{-\phi D_s}{L}(C_{eqL} - C_0), \tag{5-57}$$

where J_0 = flux at $x = 0$;
L = here the thickness of the zone of depletion;
C_{eqL} = equilibrium concentration of dissolved Fe or Mn at $x = L$;
C_0 = concentration at $x = 0$.

Again, it is assumed that a linear steady state profile between $x = 0$ and $x = L$ is maintained. For convenience we also assume that:

1. The sediment is anoxic up to the sediment-water interface and bioturbation is absent. (i.e., O_2 concentration is discontinuous at $x = 0$).

2. The layer of Fe or Mn oxyhydroxide grows upward from the sediment-water interface, i.e., the layer consists of loose flocculent particles enclosing oxygenated water.
3. Deposition of sediment from above is negligible during layer formation (this is necessary if the layer is to become monomineralic).
4. Adsorption of Fe^{++} and Mn^{++} by the anoxic sediment is negligible.
5. There are no porosity gradients, and flow of water due to compaction or externally impressed flow is negligible.

Under these conditions, by analogy with the spherically symmetric situation, we have:

$$\frac{d\ell}{dt} = \frac{\phi v_p D_s}{F_p L}(C_{eqL} - C_0), \tag{5-58}$$

$$\frac{dL}{dt} = \frac{\phi v_d D_s}{F_d L}(C_{eqL} - C_0), \tag{5-58a}$$

where ℓ = thickness of the monomineralic layer;

$\left.\begin{matrix} v_p \\ v_d \end{matrix}\right\}$ = volume of each mineral per mole of Fe or Mn.

Combining equations (5-58) and (5-58a) with (5-57), and assuming $v_d \approx v_p$, we obtain upon integration:

$$L = \left[\frac{2\phi v_d D_s}{F_d}(C_{eqL} - C_0)\right]^{1/2} t^{1/2}, \tag{5-58b}$$

$$t = \frac{\ell^2}{2H'D_s(C_{eqL} - C_0)}, \tag{5-58c}$$

where $H' = \dfrac{\phi F_d v_p^2}{F_p^2 v_d}$.

Here F_p refers to the volume fraction of Fe^{+3} or Mn^{+4} oxyhydroxide in the porous monomineralic layer (the porosity of which may be different from the value ϕ for the underlying sediment).

A quantitative examination of equations (5-58b) and (5-58c) is shown in Table 5-3 for the situation $\ell = 1$ cm, $F_p = 0.2$ (i.e., a flocculant layer), $\phi = 0.8$, $v_p \approx v_d = 20\ \text{cm}^3/\text{mole}$, $D_s = 100\ \text{cm}^2\text{yr}^{-1}$, $C_0 = 0$, and values of F_d and C_{eqL} listed in the Table. Note that even at the highest reasonable values for F_d and C_{eqL} it still takes a relatively long time for the formation of a 1-cm-thick porous layer. During this time it would be likely that the monomineralic layer would be diluted considerably by sedimented detrital particles, since times for the deposition of 1 cm of sediment range from

TABLE 5-3

Effects of varying critical parameters on the time necessary to form a 1 cm monomineralic layer of hydrous ferric or manganese oxide at the surface of a sediment by diffusion-controlled diagenetic redistribution. Assumed values: $F_p = 0.2$, $\ell = 1$ cm, $v_p = v_d = 20\ \text{cm}^3\ \text{mol}^{-1}$, $D_s = 3 \times 10^{-6}\ \text{cm}^2\ \text{sec}^{-1}$, $C_0 = 0$, $\phi = 0.8$.

F_d	% (dry wt.) dissolvable Fe or Mn	C_L μM	L (cm)	t yr
0.002	1.0	100	100	62,500
0.002	1.0	1000	100	6,250
0.004	2.0	100	50	31,250
0.004	2.0	1000	50	3,125
0.010	5.0	100	20	12,500
0.010	5.0	1000	20	1,250

about 5 years for lake and near-shore terrigenous sediments to 10,000 years for deep-sea eupelagic clay. Also, the more slowly deposited sediments generally are less likely to be anoxic (Heath et al., 1977). Thus, in order to form pure Fe and Mn layers of the type described here, one must call upon sporadic periods of unusually low sedimentation rate (sediment starvation) for anoxic sediments which otherwise are deposited at the relatively high rates typical of the lacustrine or near-shore environment.

Much diagenetic redistribution may occur via surface-reaction controlled dissolution and precipitation. In this case, simple *a priori* calculation of the time of formation of concretions and layers is impossible unless rate laws and rate constants are known for the various processes. Also, concentration gradients are no longer simple straight lines. Use of diffusion-controlled processes as shown above in the absence of water flow, does make possible calculation of an *upper limit* for rates of diagenetic redistribution. Thus, the long times necessary for the formation of monomineralic Fe-Mn layers at the sediment-water interface become even longer if surface-reaction control is operative.

Diagenetic redistribution does not occur only by the mobilization of a substance at one location and its precipitation at another. Mobilization of different components of a substance can occur at two or more different locations. The best example of this is the formation of Liesegang banding. In Liesegang banding we have the interdiffusion of two dissolved ions which can react with one another to form a relatively insoluble solid. The two ions come from different sources and when their concentrations at a given site build up, via diffusion, to sufficiently high values, precipitation of the insoluble solid occurs. This precipitation suddenly lowers concentration in the neighborhood of the solid, and as a result the diffusion profiles become altered. Continued interdiffusion results in a new build-up in concentration and precipitation at another site. Depending on the geometry of the situation this process may result in Liesegang rings (3-dimensional), tubes (2-dimensional), or layers (1-dimensional). A common example of Liesegang phenomena are rhythmic bands of iron oxides often found in sandstones. In this case precipitation is most likely brought about by the interdiffusion of dissolved Fe^{++} (from an anoxic source) and dissolved O_2 (from an oxic source). Where the Fe^{++} and O_2 meet, Liesegang banding occurs.

Mathematical treatment of Liesegang phenomena is highly complex and will not be attempted here. Much attention has been paid to this process in the laboratory by chemists studying precipitation within artificial gels. For a survey of Liesegang theories and their mathematical description, the interested reader should consult Stern (1954), Nicolas and Portnow (1973), and Flicker and Ross (1974).

Flow vs Diffusion

Discussion so far has implicitly or explicitly ignored the flow of pore water, and has treated diagenetic redistribution solely in terms of molecular diffusion. However, if there is sufficiently rapid flow, and distances over which redistribution takes place become sufficiently large, the effects of flow may become important. A simple criterion for evaluating whether or not flow is important is evaluation of the dimensionless Peclet number D_s/LU (see Lerman, 1975), where:

$$\begin{aligned} D_s &= \text{molecular diffusion coefficient;} \\ L &= \text{average distance of migration;} \\ U &= \text{interstitial water flow velocity.} \end{aligned}$$

If $D_s \gg LU$, diffusion is the major process of redistribution, and if $D_s \ll LU$, flow is the major process. In Table 5-4 are listed distances and corresponding flow rates for a typical D_s value where diffusion and flow are equally important. As can be seen, at the slow water-flow rates characteristic of mud compaction (0.0001-1 cm/yr), flow overshadows diffusion in importance only for migration distances greater than about one meter. For externally impressed flow through sands, ground water flow is sufficiently rapid ($U > 100$ cm/yr) that it should be the dominant process

TABLE 5-4

Flow rates corresponding to given distances of migration where the processes of diffusion and flow are of equal importance in diagenetic processes ($D_s = LU$). Results are for $D_s = 100\ cm^2\ yr^{-1}$ ($3.2 \times 10^{-6}\ cm^2\ sec^{-1}$).

L	$U\ (cm\ yr^{-1})$
1 mm	1000
1 cm	100
10 cm	10
1 m	1
10 m	0.1

in diagenetic transfer over all distances greater than the size of the constituent grains.

PRECIPITATION (REPLACEMENT) FROM AN EXTERNAL SOURCE

Here we discuss material precipitated in a sediment which is brought by water flow from an external source. (Of course the word "external" is not easy to define and what is really meant is that the source function for the material becomes a boundary condition. In general we may assume external sources are separated from sinks by at least several meters.) Under this heading are included both cementation and replacement, and because of its importance to early diagenesis, cementation is considered first. Two geometrical cases of cementation via water flow are discussed. They are: (1) formation (by cementation) of a spherical concretion and (2) cementation of an entire sand layer.

The situation of spherical concretion formation from flowing ground water has already been treated by the author (Berner, 1968) and detailed derivation will not be repeated here. The process is assumed to be transport-controlled and briefly, the derivation goes as follows. In the absence of flow, the concretion is assumed to grow according to the expression for the diffusion-controlled growth of a sphere (equation 5-22). With flow, the flux of material to the surface of the sphere via diffusion is supplemented by the flow term (ϕUC) (where ϕ = porosity, U = flow velocity, and C = concentration). Flow is assumed to be laminar, steady, and uniform (in other words, there are no longitudinal transverse, or time variations in U). Solution of this problem for the physically analogous case of spherical crystals falling in a fluid has been done by Nielsen (1961) and has already been presented earlier for the growth of single crystals (equation 5-25).

Integration of equation (5-25) for the conditions appropriate to concretion formation yields the expression for time of formation of the concretion:

$$t = \frac{(r_c - D_s/0.715U)(1 + r_cU/D_s)^{0.715} + D_s/0.715U}{1.715U\bar{v}(C_\infty - C_{eq})}, \qquad (5\text{-}58d)$$

where
t = time for formation;
r_c = radius of sphere;
U = ground water flow velocity;
C_∞ = (uniform) concentration at a large distance from the sphere;
C_{eq} = saturation concentration at surface of sphere;
D_s = sediment diffusion coefficient;
$\bar{v}$ = molar volume of cement.

An example of the use of equation (5-58d) for the growth of spherical calcite concretions is given in Figure 5-9.

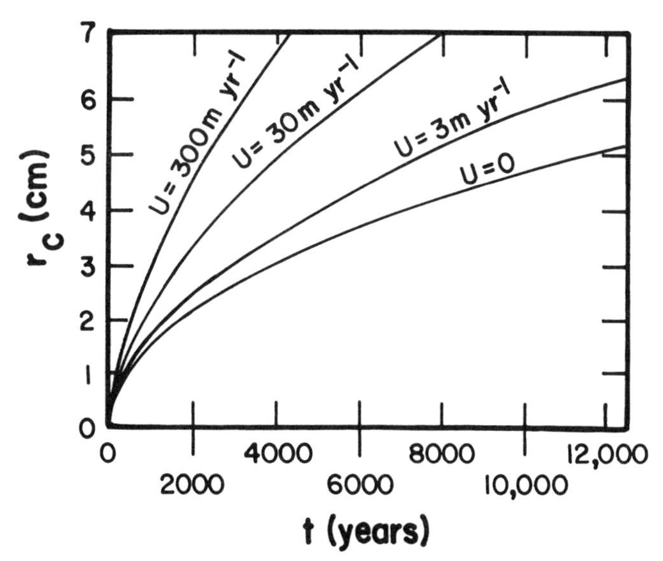

FIGURE 5-9. Minimum time necessary to form a spherical calcite concretion from flowing ground water. Plots are for transport-controlled growth and uniform, laminar flow. Assumed values: $(C_\infty - C_{eq}) = 10^{-4}M$ (10 ppm $CaCO_3$), $D_s = 10^{-5}$ cm^2 sec^{-1}, $\bar{v} = 35$ cm^3 mol^{-1}. The symbol U refers to ground water flow velocity. (After Berner, 1968.)

For the cementation of an entire sediment layer by flowing ground water, one can visualize several different geometrical situations. Here (Figure 5-10) we will consider an homogeneous sand layer sandwiched between two impermeable clay beds, through which water flows laterally with constant velocity transverse to the flow. (Boundary flow effects at the sand-clay interface are ignored.) As cement builds up near the source, porosity and permeability begin to drop and flow velocity correspondingly begins to decrease downstream. Also, molecular diffusion is assumed to be negligible compared to flow. Under these conditions the appropriate diagenetic expression is:

$$\frac{\partial \hat{C}}{\partial t} = -\frac{\partial(\hat{C}U)}{\partial y} - \frac{\partial \hat{C}}{\partial t}_{pptn}, \tag{5-59}$$

where $\hat{C}$ = concentration of dissolved material which can precipitate to form cement in mass per unit volume of total sediment;

y = horizontal distance from point of entry of pore water into the sand;

U = horizontal flow velocity;

t = time.

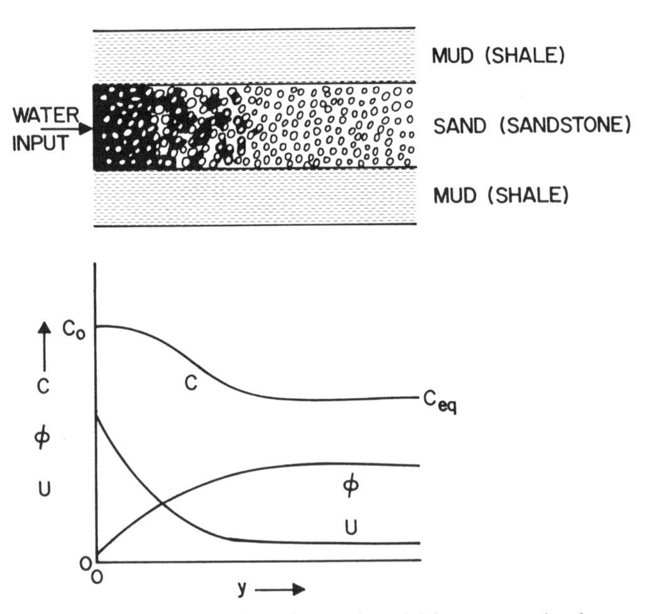

FIGURE 5-10. Schematic representation of general model for cementation by ground water flow. Cement in dark. C = concentration of cementing material in solution; ϕ = porosity; U = ground water flow velocity.

Here, the partial time derivatives refer to changes occurring at a fixed distance from the source. In terms of porosity and pore water concentration, $\hat{C} = \phi C$, so that:

$$\frac{\partial(\phi C)}{\partial t} = \frac{-\partial(\phi U C)}{\partial y} - \frac{\partial(\phi C)}{\partial t_{pptn}}. \quad (5\text{-}60).$$

Now, the change of porosity as a result of cement precipitation is given as:

$$\frac{\partial \phi}{\partial t_{pptn}} = \phi \upsilon \frac{\partial C}{\partial t_{pptn}} \quad (5\text{-}61)$$

where υ = molar volume of cementing material.

From equation (5-43) of the previous section for inhibited precipitation:

$$\frac{\partial C}{\partial t_{pptn}} = \bar{A}k(C - C_{eq})^n. \quad (5\text{-}62)$$

Substituting equations (5-61) and (5-62) in (5-60) we obtain:

$$\frac{\partial(\phi C)}{\partial t} = -\frac{\partial(\phi U C)}{\partial y} - \phi(1 + \upsilon C)\bar{A}k(C - C_{eq})^n. \quad (5\text{-}63)$$

The diagenetic equation for porosity analogous to (5-60) is:

$$\frac{\partial \phi}{\partial t} = -\frac{\partial(\phi U)}{\partial y} - \frac{\partial \phi}{\partial t_{pptn}}. \quad (5\text{-}64)$$

Since compaction is negligible in sands we can assume that flow is incompressible, i.e.:

$$\frac{\partial(\phi U)}{\partial y} = 0. \quad (5\text{-}65)$$

Substituting (5-61), (5-62), and (5-65) in (5-64):

$$\frac{\partial \phi}{\partial t} = -\phi \upsilon \bar{A}k(C - C_{eq})^n. \quad (5\text{-}66)$$

Also, multiplying (5-66) by C and subtracting the result, along with (5-65) from (5-63):

$$\frac{\partial C}{\partial t} = -U\frac{\partial C}{\partial y} - \bar{A}k(C - C_{eq})^n. \quad (5\text{-}67)$$

Thus, simultaneous solution of (5-66) and (5-67) enables one to construct plots of C and ϕ vs y and t.

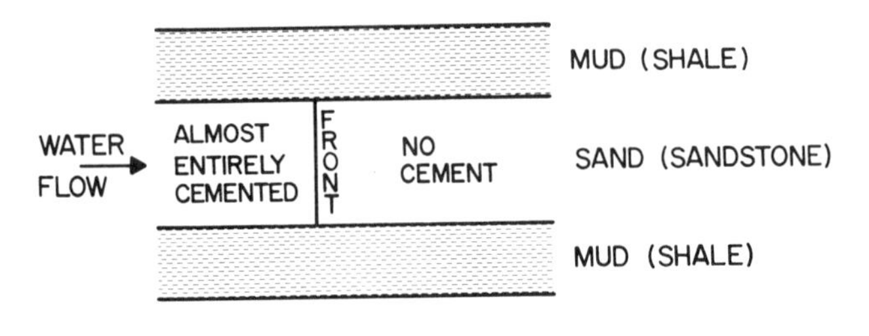

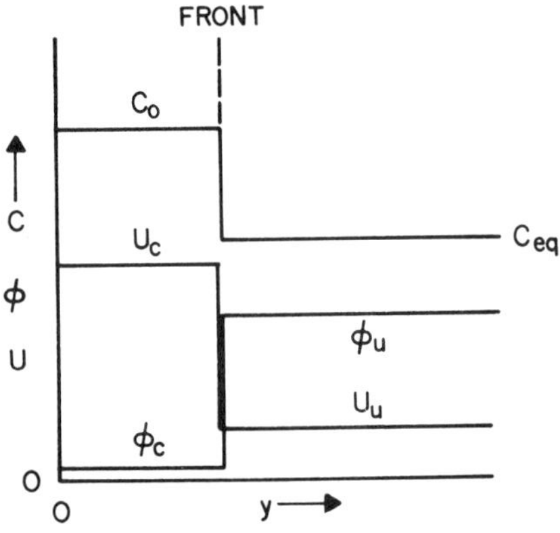

FIGURE 5-11. Schematic representation of model for cementation along an advancing front. Compare with Figure 5-10.

Because of problems in obtaining values of k, no attempt will be made here to solve these equations. Instead we will consider a special case where precipitation is confined to a zone which is very narrow compared to the thickness of fully cemented rock. In other words, one can visualize cementation as proceeding along a front where a very sharp gradient in concentration occurs (see Figure 5-11). Beyond the front, water is saturated with the precipitating material and no cementation can take place. Behind the front, cementation is essentially complete, leaving only a little residual (or uncementable) porosity to provide sufficient permeability for water to flow through the cemented zone. Such residual permeability is necessary or cementation would always be self-limiting and could never continue. (Decrease in rate of cementation due to permeability decrease is emphasized by Pettijohn et al., 1972).

Mathematical treatment of the cementation front is by means of an infinitely thin box model for the frontal zone. In other words, we treat the front as a discontinuity. At this discontinuity we have, at steady state, a balance between input, and precipitation. In other words:

$$Q(C_0 - C_{eq}) = J_{pptn}, \tag{5-68}$$

where C_0 = concentration of cementing material in the pore water behind the front;
C_{eq} = concentration in pore water beyond the front, equal to the saturation value;
J_{pptn} = rate of precipitation per unit area of front in mass per area per time;
Q = flux of water through the front in volume of water per unit area per unit time.

As a result of the addition of cement the front advances at the rate:

$$U_F = \left(\frac{v}{\phi_u - \phi_c}\right) J_{pptn}. \tag{5-69}$$

where U_F = velocity of migration of the front away from the water source;
v = molar volume of cementing material;
ϕ_u = porosity of the uncemented zone beyond the front;
ϕ_c = porosity of the cemented zone behind the front ($\phi_u - \phi_c$ represents the volume of cement per volume of total sediment).

Combining (5-68) and (5-69):

$$U_F = \left(\frac{vQ}{\phi_u - \phi_c}\right) (C_0 - C_{eq}). \tag{5-70}$$

In general $(\phi_u - \phi_c) \approx \phi_u$, or in other words, porosity of the cemented rock is negligibly low. Using this approximation, and introducing the definition:

$$U_u = Q/\phi_u, \tag{5-71}$$

where U_u = velocity of water flow in the uncemented zone beyond the front, we obtain, finally:

$$U_F \approx v(C_0 - C_{eq})U_u, \tag{5-72}$$

which can be expressed, alternatively, since $U_u\phi_u = U_c\phi_c$, as:

$$U_F \approx v\frac{\phi_c}{\phi_u}(C_0 - C_{eq})U_c, \tag{5-73}$$

where U_c = velocity of water flow in the cemented zone.

These expressions thus give the rate of migration of the cementation front (U_F) away from its source in terms of the rate of water flow to the front (U_c) (equation 5-73) or rate of flow away from the front (U_u) (equation 5-72).

An idea of the order of magnitude of U_F and the applicability of the cementation front model can be gained by considering the situation of cementation by calcium carbonate. For calcite in ground water a reasonable value of $(C_0 - C_{eq})$ is 10^{-7} moles per cm^3 (10 ppm $CaCO_3$). A typical ground water flow velocity for well-cemented sandstone U_c is 10 meters per year. From these values we have, using equation (5-73) and assuming $\phi_c/\phi_u = 0.1$, the result $U_F = 0.35$ cm per thousand years. To cement a sand body laterally along a front over distances of a few kilometers would, thus, require periods on the order of a billion years. Geological evidence shows that cementation occurs much more rapidly than this. Thus, if cement comes from an external source, the water flow rate that we have used (10 meters per year) must be considerably too low. Higher flow rates imply higher porosities (actually higher permeabilities) which in turn means that rapid precipitation along a sharp front, with formation of a fully cemented, impermeable zone behind the front through which water must pass, is not a good model for cementation over large distance (>1 km). Rather it appears that regional cementation must occur diffusely over a wide zone because of highly inhibited, surface-reaction controlled precipitation. (This can be visualized as the coalescence of "concretions.") Either this, or much of the cement during large-scale cementation is derived internally by diagenetic redistribution.

From equations (5-72) or (5-73) it can also be shown that at least 300,000 volumes of pore water must pass through a given volume of sand to enable the pore space to be completely filled with calcite. This result depends only upon the degree of supersaturation of the water and the molar volume of calcite (and, thus, is model independent) and shows that large volumes of water are needed to bring about appreciable cementation.

Replacement by material brought in from an external source often occurs on a large scale. Because of this we will here treat only the problem of massive replacement of specific minerals in an entire sedimentary layer. Replacement on the scale of a layer can be visualized in terms of essentially the same model as that presented above for the cementation of a permeable sand layer sandwiched between two impermeable clay layers and subject to laterally uniform water flow. The only difference is that the precipitated matter replaces pre-existing grains and does not fill pore space.

Equation (5-60) can thus also be used for replacement:

$$\frac{\partial(\phi C)}{\partial t} = -\frac{\partial(\phi U C)}{\partial y} - \frac{\partial(\phi C)}{\partial t_{pptn}}.$$

Because replacement does not involve porosity changes, we can assume $\partial \phi / \partial t = 0$. Also, if porosity is uniform in the sand layer, $\partial \phi / \partial y = 0$ and as a result $\partial U / \partial y = 0$. Under these conditions equation (5-60) reduces to:

$$\frac{\partial C}{\partial t} = -U \frac{\partial C}{\partial y} - \frac{\partial C}{\partial t_{pptn}}. \tag{5-74}$$

Substituting equation (5-62) for the rate of precipitation we obtain:

$$\frac{\partial C}{\partial t} = -U \frac{\partial C}{\partial y} - \bar{A}k(C - C_{eq})^n. \tag{5-75}$$

This is the basic expression which, upon integration, enables us to construct theoretical plots of C vs y at various times of replacement.

As was the case for cementation, we will not attempt to solve (5-75), but rather will consider the simplified situation of replacement along a front, analogous to cementation along a front as represented by Figure 5-11. The situation is much simpler than for cementation in that porosity does not change as a result of replacement. The replacement front can be visualized as separating completely replaced grains from completely unreplaced grains. The rate of advancement of the front, by analogy with (5-69), is given by:

$$U_F = \left(\frac{\upsilon}{\phi_g}\right) J_{pptn}, \tag{5-76}$$

where ϕ_g represents the volume fraction represented by the grains which undergo replacement (ϕ_g generally does not include all the sediment grains, i.e., $\phi_g \neq (1 - \phi)$) and the other symbols are defined in equations (5-68) and (5-69).

Combining equation (5-76) with (5-68) we obtain:

$$U_F = \left(\frac{\upsilon Q}{\phi_g}\right)(C_0 - C_{eq}). \tag{5-77}$$

Now, the lateral flow velocity U is given by:

$$U = Q/\phi. \tag{5-78}$$

Introducing the definition:

$$\psi = \phi_g/\phi v \quad (5\text{-}79)$$

where ψ is the mass of replacement material per unit volume of pore water, we obtain finally:

$$U_F = \frac{(C_0 - C_{eq})}{\psi} U. \quad (5\text{-}80)$$

Equation (5-80), which has been independently derived by Verigin (see Dobrovolsky and Lyalko, 1979), thus expresses the velocity of travel of the replacement front U_F in terms of the velocity of water flow U. This equation is independent of the kinetics of precipitation and rests only on the assumption of a sharp front of replacement. It may apply to many natural situations.

Causative Factors in Authigenic Processes

Much has been said so far about how and how fast authigenic processes occur, but the question of why they occur has been largely unaddressed. Cementation, replacement, or diagenetic redistribution all result in a decrease in Gibbs free energy, and it is an excess of free energy which provides the driving force for authigenesis. Some of the excess free energy is provided initially by thermodynamically unstable minerals and mineraloids such as aragonite, opaline silica, volcanic glass, limonitic goethite, and semi-amorphous clay minerals. These substances are all less stable than corresponding authigenic minerals (e.g., calcite, quartz, hematite, kaolinite, etc.) regardless of the composition of the surrounding pore water. Another important causative factor is excess free energy brought about by a diagenetic change in pore water composition. Alteration of pore water composition can be brought about simply by dissolution of unstable substances such as those cited above (water equilibrated with opaline silica is supersaturated with respect to quartz), but an important additional cause of altered composition is the bacterial decomposition of organic matter. Reactions involving organic substances can bring about drastic changes in the composition and redox state of interstitial waters during early diagenesis, which in turn causes initially deposited stable minerals to become unstable. A good example is the formation of pyrite by the reaction of hematite with biogenically-derived H_2S (Goldhaber and Kaplan, 1974; Rickard, 1975).

A complete discussion of causative factors in authigenesis requires detailed coverage of many broad topics such as the origin of limestone, dolostone, chert, pyrite, etc. This is beyond the scope of the present book and the interested reader is referred instead to standard works on sedimentary petrology and geochemistry (e.g., Engelhardt, 1977; Füchtbauer, 1974; Lippmann, 1973; Bathurst, 1971; Pettijohn, 1975; Blatt et al., 1972; Berner, 1971). Our goal here will be simply to treat a few subjects which are readily amenable to theoretical explanation and modeling. Subjects discussed are: nucleation and crystal growth during cementation, and the origin of cement-type mineral segregations (concretions, layers, etc.) especially as they relate to fossilization.

Cementation (the filling of pore space with authigenic material) is a common process whereby loose sand, gravel, or the like is converted into solid rock. The common cementing substances are calcite, quartz, and various iron minerals (siderite, hematite, pyrite) which are also the common minerals of concretions. Cementation comes about when a supersaturated interstitial water contacts an appropriate surface which promotes heterogeneous nucleation of the cementing material on the host detrital grains. Sediments are an especially good nucleating medium because of the variety and abundance of nuclei per unit volume of pore water. In addition, the contact between grains provides an unusually favorable situation for nucleation and, in fact, this is often where initial cementation normally begins (e.g., see Bricker, 1971). The reason why grain contacts are favored can be seen from Figure 5-12 and the following argument (based on the model of Wollast, 1971): If one greatly enlarges the zone of grain contact, it can be thought of as a very narrow parallel-sided "crack." The free energy of formation of a single disc-shaped nucleus

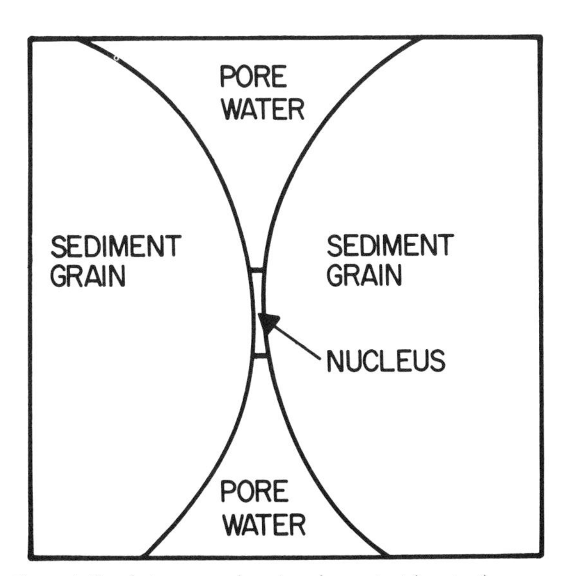

FIGURE 5-12. Cementation between grains at grain contact (see text).

in such a "crack" (by analogy with equation (5-5)) is:

$$\Delta F^* = -n^* k_B T \ln \Omega + 2(\sigma_s - \sigma_w)A + \sigma_n h S, \quad (5\text{-}81)$$

where σ_w = specific interfacial (surface) free energy between the material of the host grains and the interstitial water;

σ_s = specific interfacial free energy between the material of the host grains and the nucleus;

σ_n = specific interfacial free energy between the nucleus and water;

$2A$ = surface area of flat sides of disc;

h = thickness of the disc;

S = circumferance of the disc.

As the disc becomes thinner and thinner or, in other words, the grains constituting the "walls" get closer and closer together, the last term in equation (5-81) becomes less and less important. As $h \to 0$ then we have:

$$\Delta F^* \approx -n^* k_B T \ln \Omega + 2(\sigma_s - \sigma_w)A. \quad (5\text{-}82)$$

In this limiting situation we are presented with a most interesting result. For the host grains to be reasonably good nucleating agents, $\sigma_s < \sigma_w$. This causes the second term in equation (5-82) to be negative. Thus, as the difference $(\sigma_s - \sigma_w)$ increases, less and less free energy is required to nucleate the cementing material. In fact if $(\sigma_s - \sigma_w)$ becomes sufficiently large, it is at least theoretically possible to form the disc-shaped nucleus from an undersaturated solution. At any rate, equation (5-82) shows why nucleation at contacts between grains is favored.

As sediments are buried, grain contacts become unfavorable loci for cementation. This is because the grain contacts become points of excess pressure due to the weight of the overlying sediment particles. This gives rise to pressure solution at grain contacts in response to super-solubility caused by the excess pressure. In this case material migrates away from the grain contacts and precipitates in the open portion of the pore water where pressures are hydrostatic, and, therefore, lower. Several theoretical models have been proposed to explain pressure solution (Weyl, 1959; deBoer, 1977; Robin, 1978). However, they will not be discussed here since the topic of pressure solution belongs more properly under a discussion of deep burial phenomena which is beyond the scope of this book.

One might expect that certain host grains would serve as better nucleating agents, for a given mineral, than others. For example, one would expect cementation of a quartzose sand by calcite to be initiated on calcitic shell fragments scattered within the sand. Growth around the shell fragments would bring about the formation of concretions which would eventually coalesce to bring about complete cementation. Studies of the nuclei of calcareous concretions, however, do not generally bear this out. It is true that some concretions enclose delicate calcareous fossils and preserve them from further diagenetic alteration (e.g., Waage, 1964). However, many other concretions form about non-calcareous nuclei such as fossil insects, fish, and leaves so that it is not entirely clear whether precipitation of the calcium carbonate in concretions is brought about by the calcareous or by the organic parts of shelled organisms. In fact, most concretions have no obvious nuclei (Twenhofel, 1961). In recent marine sediments, calcareous cements, which consist solely of aragonite and highly magnesian calcite, do not form selectively on host grains of the same mineralogy as might be expected. Magnesian calcite cement forms equally well on detrital grains of either aragonite or high-magnesian calcite (Bricker, 1971).

Alterations in surface chemistry brought about by the adsorption of trace inhibitors may explain why certain detrital grains do not behave as good nucleating agents for cement of the same mineralogy. Several studies have shown that dissolved humic substances and orthophosphate can act to inhibit severely the precipitation of calcium carbonate both in the laboratory (Simkiss, 1964; Chave and Suess, 1970; Berner et al., 1978) and in the pore waters of modern organic-rich sediments (Berner et al., 1978). This results in pore waters highly supersaturated with respect to both calcite and aragonite occurring in contact with many scattered shell fragments. Probably the surfaces of the shell fragments are poisoned toward further $CaCO_3$ precipitation. In this case, any grain may be as good a nucleating agent for $CaCO_3$ precipitation as $CaCO_3$ itself.

Besides nucleation and crystal growth, another important factor in cementation is water flow. If cementing material is not derived from local diagenetic redistribution, as is often the case, it must come from outside the rock. The amount of potential cement carried by a given volume of pore water is extremely small compared to the volume of pore space to be filled. From our previous calculation we found that at least 300,000 volumes of pore water must flow through a given volume of pore space in order to fill it completely with calcite cement. This requires an appreciable flow of water. One of the reasons sandstones seem to be more readily cemented than shales may be that the greater permeability of sands allows a greater volume of water to flow through them over a given amount of time.

A causative factor in the formation of $CaCO_3$ concretions about fossil organisms, as alluded to above, may be the bacterial decomposition of the organic materials left upon death of the organism. Decomposition of biological soft tissue under anoxic conditions often leads to a significant rise in carbonate alkalinity (e.g., Berner, 1969a). This rise is due to ammonia and CO_2 liberated by the biological decomposition of proteins and their constituent amino acids:

$$\text{organic C} \rightarrow CO_2,$$
$$\text{organic N} \rightarrow NH_3,$$
$$2NH_3 + CO_2 + H_2O \leftrightarrows 2NH_4^+ + CO_3^{--}.$$

If the sediment is open to flow by sulfate-rich solutions such as seawater, additional alkalinity can be generated as a consequence of bacterial sulfate reduction:

$$2CH_2O + SO_4^{--} \rightarrow H_2S + 2HCO_3^-.$$

The overall effect is to raise the concentration of carbonate ion and, in turn, to raise the degree of supersaturation with respect to $CaCO_3$. As a result, precipitation of $CaCO_3$ can occur around the decomposing protoplasm. Once the organic matter is thoroughly decomposed or sulfate becomes unavailable, the source of carbonate is cut off and precipitation ceases. The continued growth of a calcareous concretion about a soft-bodied organism, then, must be due to supersaturation brought about by external factors. However, the initial $CaCO_3$ precipitated by biogenic processes may serve as nuclei for continued concretion growth.

The initial calcium-containing precipitate brought about by ammonia generation need not be $CaCO_3$, and in some cases, e.g., fish and humans, it is not. Instead calcium soaps derived from the fat of the organism can occur, forming a natural substance known as adipocire (Bergmann, 1963). The writer has demonstrated how fish adipocire may form from seawater in the laboratory (Berner, 1969a). The steps in adipocire formation can be summarized as follows:

$$\text{Protein N} \rightarrow NH_3$$
$$\text{Fats} \rightarrow \text{fatty acids (RCOOH)}$$
$$NH_3 + RCOOH \rightarrow NH_4^+ + RCOO^-$$
$$Ca^{++} + 2COO^- \rightarrow Ca(RCOO)_2,$$

where R = hydrocarbon chain. Once formed, the $Ca(RCOO)_2$ or adipocire is thermodynamically unstable relative to $CaCO_3$ plus hydrocarbons and may break down with time to form $CaCO_3$. This $CaCO_3$ can then act, as in the case above for initially formed $CaCO_3$, as nuclei for localizing further growth around the fossilized organisms.

The bacterial decomposition of organic matter is probably more important in forming concretions consisting of authigenic iron minerals than for those consisting of calcium carbonate. An especially good example is provided by pyrite (Berner, 1969b). During early diagenesis pyrite forms by the reaction of hydrogen sulfide with reactive detrital iron minerals. The hydrogen sulfide in turn, in marine sediments, comes from the reduction of dissolved sulfate (see Chapter 6) by bacteria which use decomposing organic matter as a reducing agent and an energy source. Thus, if we have an originally heterogeneous distribution of organic matter in a sediment, we would expect various centers of bacterial sulfate reduction. At each clot of organic matter hydrogen sulfide would be produced which would rapidly precipitate any nearby dissolved iron to form various insoluble iron sulfides. These sulfides would eventually be transformed to pyrite. Because of sulfate reduction and iron sulfide precipitation, lowered concentrations of dissolved sulfate and Fe^{++} would occur in the vicinity of the organic clots. If the sediment is more or less homogeneous and anoxic (but not sulfidic) between the clots, we would expect to find spherically symmetrical diffusion of both SO_4^{--} and Fe^{++} towards the clots. Supply of additional sulfate could come from the diffusion (plus bioturbational irrigation) of seawater sulfate into the sediment, whereas additional dissolved Fe^{++} could come from the reduction of disseminated detrital iron oxyhydroxides. This is all illustrated in Figure 5-13. With continued

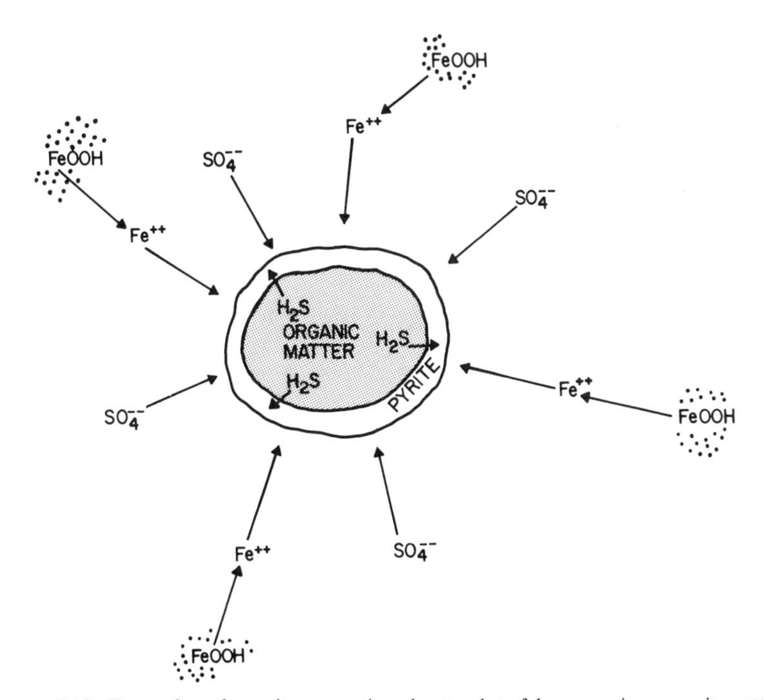

FIGURE 5-13. Formation of a pyrite concretion about a clot of decomposing organic matter.

diffusion of SO_4^{--} and Fe^{++}, reduction of SO_4^{--} by organic matter decomposition, and precipitation of iron sulfides, a build-up of pyrite would occur adjacent to the organic matter. In this way early diagenetic pyrite concretions and pyritized fossils might form. The scale of the concretions can vary from micron-sized "framboids," to millimeter-sized fillings of foraminifera tests, to large meter-long pyrite bodies (from dead whales?).

The amount of pyrite present and therefore the size (assuming sufficient iron and dissolved sulfate in the surrounding sediment) is limited by the amount of decomposable organic matter in the clot. The rate of growth of such a concretion of pyrite, if controlled by diffusion, can be calculated from equation (5-58) given earlier.

Pyrite-rich layers can form by a similar mechanism. Assuming that diffusion gradients of dissolved SO_4^{--} and Fe^{++} are set up around a layer of decomposing organic matter, one can also calculate the time of formation of a monomineralic pyrite layer by diffusion (Berner, 1969b). If the process is diffusion-controlled, the proper growth expression is equation (5-58c).

Whether or not an early diagenetic pyrite concretion or layer forms depends not only upon an heterogeneous distribution of organic matter but also upon the concentration of *reactive* iron in the sediment. (All iron in the sediment is not reducible or reactive toward H_2S.) If the iron content is sufficiently low, hydrogen sulfide can build up and diffuse away from the clot or layer of organic matter. In this way pyrite precipitation is delocalized and migration of iron to form mineral segregations does not occur. On the other hand, the rate at which the H_2S-rich front advances away from its source and converts detrital iron minerals to pyrite can be calculated (Berner, 1969b) by expressions similar to that for replacement along a moving front, equation (5-80).

Part II

APPLICATIONS

6

Marine Sediments of the Continental Margins

Sediments deposited near the continents are discussed in this chapter. This includes a wide variety of environments ranging from the upper continental slope and outer shelf to near-shore deltas, bays, etc. (Brackish or hypersaline lagoons and estuaries are considered separately in Chapter 8.) The principal features that distinguish these sediments from those discussed in the succeeding chapters is that they are deposited at moderately rapid to rapid rates (0.01 to 1 cm per year) in waters of normal marine salinity (roughly 35 $\pm$ 5‰). Only fine-grained sediments are discussed because they exhibit greater diagenetic variation with depth, are more easily sampled by coring, and have been studied from a diagenetic standpoint much more so than sands.

Associated with higher sedimentation rates one also encounters higher organic matter contents in fine-grained continental margin sediments than in deep-sea sediments. Heath et al. (1977) have shown that a reasonably good positive correlation exists between rate of deposition and organic carbon content of marine sediments in general. A higher organic content is important because most of the early diagenetic changes exhibited by continental margin sediments are due, either directly or indirectly, to the microbial decomposition of organic matter (Berner, 1974). These sediments are enriched in organics owing to two factors. First, biological production (by plankton) in the overlying water is higher nearer shore because of an increased input of nutrients from rivers or from coastal upwelling. Second, there is more organic matter in rapidly deposited sediments, because less destruction by organisms living at the sediment-water interface has occurred prior to burial. In pelagic sediments, by contrast, hundreds to thousands of years are available for decomposition before burial and, as a result, the sediments are very low in organic matter.

Some of the chemical changes accompanying organic matter decomposition have been briefly outlined in Chapter 4 in the discussion of microbial reactions. These include depletion and destruction of organic compounds (with consequent release of phosphate and ammonia), deoxygenation, reduction of NO_3^- to N_2, reduction of MnO_2 to Mn^{++} and Fe_2O_3 to Fe^{++}, reduction of SO_4^{--} to H_2S, and the formation of methane. Reaction of these products with one another or with other constituents results in the formation of authigenic minerals, namely pyrite (FeS_2) from reaction of H_2S with Fe^{++}, rhodocrosite ($MnCO_3$) from reaction of HCO_3^- (from sulfate reduction) with Mn^{++}, calcite and aragonite from reaction of HCO_3^- with Ca^{++}, and apatite ($Ca_5(PO_4)_3(OH,F)$) from reaction of PO_4^{--} with calcite. In addition, exchange reactions may occur such as the uptake by smectite of Mg^{++} to replace Fe^{++} lost by pyrite formation (Drever, 1974), or the exchange of common cations for one another on new surfaces coated with organic compounds (Manheim, 1976). In this chapter we will not attempt to describe all of these processes; rather, selected studies will be discussed which have approached organic matter decomposition in continental margin sediments from a theoretical standpoint. The literature in this field is extensive, and the interested reader is referred to the compilations of Shishkina (1972), Glasby (1973), and Manheim (1976) for material not covered here.

Our treatment of early diagenesis in continental margin sediments is divided into two sections; diagenesis within the zone of bioturbation and diagenesis below the zone of bioturbation. This approach stresses the importance of bioturbation as it affects those processes associated with the microbial decomposition of organic matter. The "zone of bioturbation" is defined here as including all sediment depths which are appreciably affected by mixing with the overlying water. Since only sediments overlain by aerated seawater undergo bioturbation (Rhoads and Morse, 1971), we are concerned, thus, only with those depths which undergo (at least) periodic oxygenation. Sediments deposited in anoxic bottom water (e.g., the Black Sea), where there is no bioturbation, and those buried to sufficient sediment depths are considered to be below the zone of bioturbation. The boundary between bioturbated and non-bioturbated sediment depends on sediment type, and ranges from a few centimeters to one meter or more.

Diagenesis Within the Zone of Bioturbation

Theoretical description of diagenesis in the zone of bioturbation is exceedingly difficult. Not only are solute migration and particle transport altered by burrowing organisms, but also chemical reactions change in rate and type of reaction over very short distances due to sharp compositional gradients and with time due to seasonal temperature fluctuations. In addition, sporadic disturbance by wave and current stirring (in shallow water) may occur, further complicating the picture. The theoretical approaches presented here are only first attempts at describing, quantitatively, the multifaceted aspects of early diagenesis within the zone of bioturbation.

All of the organic reactions cited above may occur simultaneously within the zone of bioturbation. This is so because of strong horizontal gradients in organic matter content and interstitial water composition. Strong gradients in organic matter result from the presence of discrete organic remains within shells and tests or as "dead bodies." Horizontal gradients in pore water composition are brought about by stirring and by irrigation of burrows with overlying seawater by sedentary infauna. The normal depth succession of deoxygenation, denitrification, sulfate reduction, and methane formation can occur over short distances as one goes, for example, from irrigated burrows to a nearby snail shell containing the dead remains of its original inhabitant. Because of these strong gradients sampling of sediments in the zone of bioturbation is difficult, and must always involve some averaging of properties. To avoid the problems of "microenvironments," discussion here will be restricted to average properties of sediments within this zone. This means that we will consider only data obtained by lateral homogenization of sediment over several tens to hundreds of square centimeters for each depth sampled. This procedure is sufficient to smooth out effectively the heterogeneity due to burrows, corpses, and other factors, and enables one to focus on average changes occurring with depth (e.g., Hanor and Marshall, 1971).

Aller (1977; 1980) has presented considerable data for diagenetic changes occurring within the zone of bioturbation of sediments from Long Island Sound, U.S.A. Only his results for ammonia, which exemplify his theoretical approach, will be discussed here. Ammonia in these sediments is produced by the anoxic decomposition of organic nitrogen compounds. The ammonia undergoes oxidation (by bacteria) to nitrite and nitrate upon migration into oxygenated water contained in nearby burrows or contained in the constantly agitated (and, therefore, oxygenated) top centimeter of sediment. Between burrows the ammonia builds up in otherwise anoxic sediments and undergoes adsorption, as NH_4^+, on the surfaces of the sediment particles. Because of seasonal temperature changes in the near-surface sediments the rates of bioturbation and bacterial ammonia formation change. This gives rise to corresponding changes in the profiles of ammonia concentration-vs-depth over the year. Actual measurements by Aller for the station NWC are shown in Figure 6-1.

Aller has attempted to explain the fluctuating profiles shown in Figure 6-1 in terms of time-dependent, that is, non-steady state, diagenetic modeling. Aller's time-dependent model rests on the following assumptions:

1. A one-dimensional depth model is adopted with the sediment-water interface as origin.

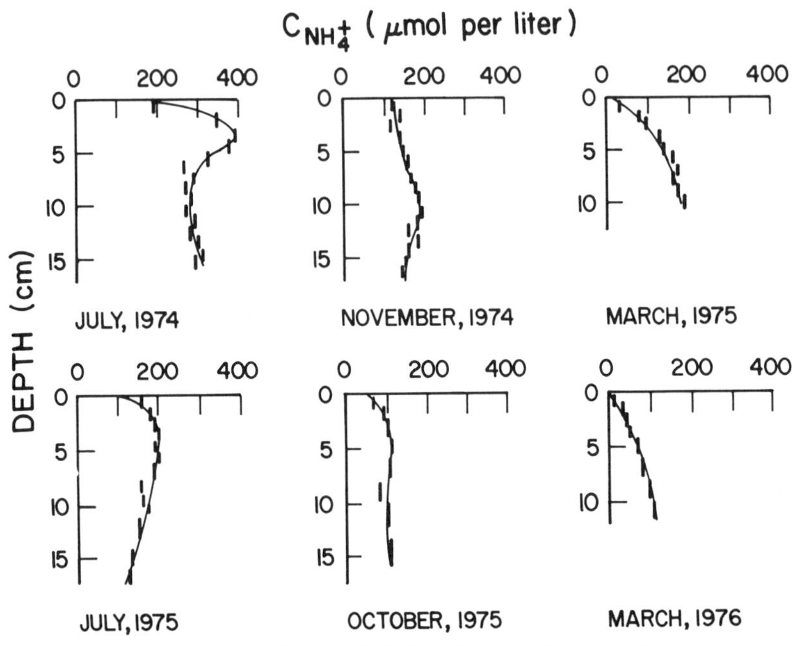

FIGURE 6-1. Concentration of dissolved NH_4^+ vs depth as a function of time in sediments at Long Island Sound site NWC. (After Aller, 1977; 1980.)

2. Organic nitrogen decomposition occurs at the rates and temperature and depth dependence measured on the same sediments in the laboratory. (Depth dependence is exponential.)
3. Burial advection can be ignored over the time scale of one year (i.e., one full temperature cycle).
4. Bioturbation can be described in terms of a one-dimensional biodiffusion coefficient for the pore water which combines the effects of irrigation and particle mixing ($D_B + D_I$).
5. Adsorption is rapidly reversible and follows a simple linear isotherm.
6. Oxidation of ammonia is negligible.
7. Compaction, water flow, porosity gradients, etc. are ignored.

Aller determined that the temperature dependence of ammonia formation has a Q_{10} value of 3. In other words, the rate is increased by a factor of 3 for a ten-degree increase in temperature. For bioturbation Aller assumed $Q_{10} = 2$, which means the biodiffusion coefficient increases by a factor of 2 with a ten-degree increase in temperature. These values are based on actual measurements. Seasonal changes in rates, due to temperature change, were described by expressions of the form:

$$g(t) = g_0 \exp\left[\frac{-A}{B + Z \sin\left(\frac{2\pi}{\alpha} t\right)}\right], \tag{6-1}$$

where A, B, Z = constants (A is related to Q_{10});

α = frequency of oscillation in cycles per year ($\alpha = 1$ for annual variation);

$g(t)$ = seasonal variation function.

From the above assumptions the proper diagenetic expression from (3-74) is:

$$\frac{\partial C}{\partial t} = \left(\frac{D_{BI_2} g_1(t)}{1 + K_N}\right)\frac{\partial^2 C}{\partial x^2} + \frac{R_0 \exp(-\varepsilon x) g_2(t)}{1 + K_N}, \tag{6-2}$$

where D_{BI_2} = average biodiffusion coefficient for dissolved ammonia at 22°C in this sediment. ($D_B + D_I$);

R_0 = rate of production of dissolved ammonia at the sediment-water interface and at 22°C;

$g_1(t), g_2(t)$ = expressions of the form (6-1) for biodiffusion and ammonia production respectively;

ε = empirical constant.

Solution of this equation was done analytically using the initial condition:

$$t = 0,$$
$$C = a_0 + a_1 x \ (a_0, a_1 = \text{constants}),$$

and upper boundary condition (at depth the sediment was treated as a semi-infinite solid):

$$t > 0,$$
$$x = 0,$$
$$C = 0,$$

to give a very complex expression which will not be repeated here (see Aller, 1977, for details). Plots of this expression for different periods of time are shown in Figure 6-2. As can be seen, fluctuations similar to those actually found are observed, with a concentration maximum occurring

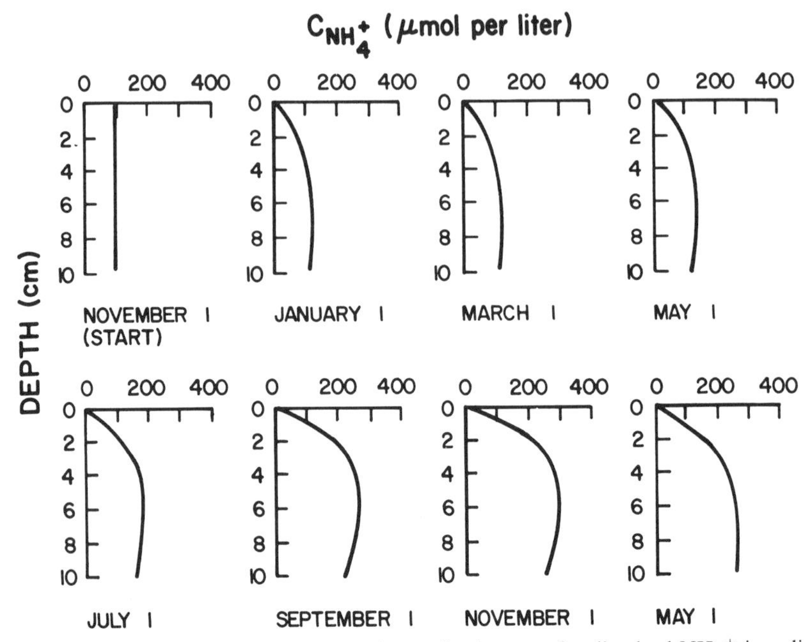

FIGURE 6-2. Model-calculated concentration-vs-depth curves for dissolved NH_4^+ in sediments from site NWC of Long Island Sound. Model is one-dimensional and considers sinusoidal variations of D_{BI} and R with time. (After Aller, 1977.) Compare with Figure 6-1.

during certain times of the year. A more exact reproduction by the model of the actual curves might have been obtained if the following factors had been considered: the depth dependence of $g(t)$ due to temperature gradients, the depth dependence of D_{BI_2} (which is instead assumed to be a constant over the zone of bioturbation for a given temperature), enhanced ammonia production rates resulting from the settling out of fresh planktonic material following plankton blooms, and use of a different model to describe bioturbation.

Aller (1977; 1980) has evaluated the last factor by constructing an alternative irrigation model. The ammonia was assumed to be removed from the sediment by molecular diffusion into burrows which are constantly flushed with overlying oxygenated seawater. The burrows are vertical cylinders and diffusion into them by ammonia occurs laterally with radial symmetry. Figure 6-3 illustrates the overall geometry. A brief discussion of this model has already been given in Chapter 3. The appropriate steady state diagenetic equation which expresses both radial and

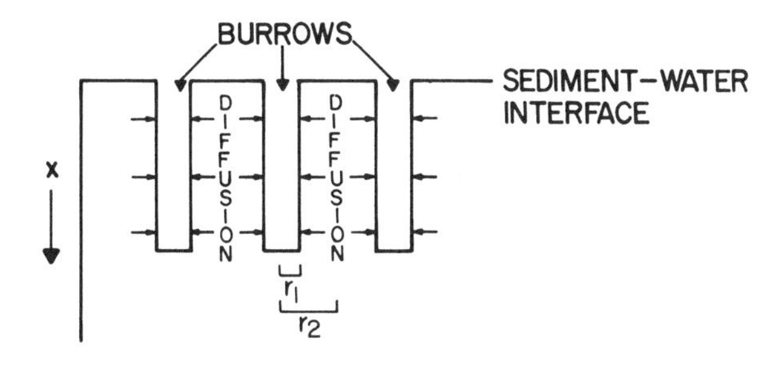

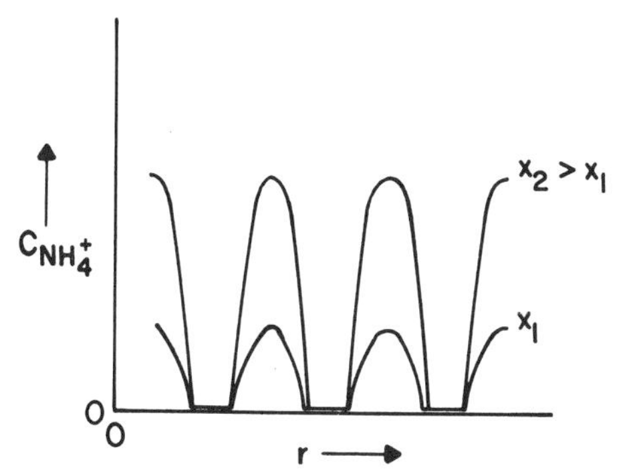

FIGURE 6-3. Simplified version of the cylindrical-burrow irrigation model of Aller (1977; 1980). Note zero concentration of NH_4^+ within burrows and maximum concentration between burrows.

vertical diffusion is:

$$\frac{\partial C}{\partial t} = \frac{D_s}{r}\frac{\partial\left(r\dfrac{\partial C}{\partial r}\right)}{\partial r} + D_s\frac{\partial^2 C}{\partial x^2} - \omega\frac{\partial C}{\partial x} + R, \qquad (6\text{-}3)$$

where r = radial distance from the axis of the nearest burrow (see Figure 6-3);

R = rate of ammonia production.

Aller showed that at the burrow sizes and densities found in most sediments, steady state is rapidly re-established relative to seasonal temperature changes. Also, at the sedimentation rates and sediment depths considered, it can be shown that the burial advection term is small in comparison with the other terms of equation (6-3). Thus, equation (6-3) was reduced to:

$$\frac{D_s}{r} \frac{\partial\left(r \frac{\partial C}{\partial r}\right)}{\partial r} + D_s \frac{\partial^2 C}{\partial x^2} + R = 0. \tag{6-4}$$

The appropriate boundary conditions for the solution of (6-4) are that the concentration of ammonium within each burrow, due to flushing (and oxidation), is a constant, low value. In other words, where $r = r_1$ (the radius of the burrow), $C = C_0$ (value for the overlying seawater). At the "divides" between burrows (see Figure 6-3) ammonium concentration is assumed to be at a maximum such that $(\partial C/\partial r)_{r=r_2} = 0$. (Here $2r_2$ represents the distance between burrows.) Finally, across the bottom of the zone of irrigation vertical flux is assumed equal to that calculated from Fick's Law ($= \phi D_s \partial C/\partial x$).

Using these boundary conditions Aller (1980) solved equation (6-4) to give a rather lengthy analytical expression for C as a function of x and r. To compare theory with measurements, from this expression he calculated the concentration for a given depth interval averaged over many burrows. This is in keeping with his sampling procedure which involved homogenization of mud samples taken over a large area by means of box coring. The appropriate expression for average concentration $\bar{C}$ is:

$$\bar{C} = \frac{2\pi \int_{x_1}^{x_2} \int_{r_1}^{r_2} Cr\, dr\, dx}{2\pi \int_{x_1}^{x_2} \int_{r_1}^{r_2} r\, dr\, dx}. \tag{6-5}$$

By using measured values of r_1 and choosing an appropriate value of r_2 (treating r_2 as a curve-fit parameter) Aller was able to calculate curves of $\bar{C}$ vs x for different times of the year. He found that he could closely fit the data of Figure 6-1 by using a value of r_2 not much different from that calculated from the measured density of burrowing infauna. Thus, he concluded that the cylindrical burrow model could be used to predict concentration profiles and to explain high degrees of solute loss by irrigation without invoking crude, *a priori* parameters such as D_{BI}. All that is needed is molecular diffusion and an appropriate geometrical description of the sediment-water interface, which in this case is treated as being highly invaginated (see Figure 6-3). Unfortunately, this approach involves greater mathematical complexity, but this only reflects the greater complexity of diagenesis within the zone of bioturbation.

Mixing of sediments near the sediment-water interface need not occur solely as a result of bioturbation. The study of Vanderborght et al. (1977a) has shown that extensive mixing by waves and currents takes place in the upper 3.5 centimeters of the organic-rich, fine-grained muds near the Belgian coast of the North Sea. These sediments exhibit a low activity of benthic macrofauna and meiofauna probably due to the constant resuspension of sediment. A physical mixing coefficient (analogous to the biodiffusion coefficient) has been calculated from diagenetic modeling of pore water data for dissolved silica. The diagenetic equation used is:

$$D(x)\frac{\partial^2 C}{\partial x^2} - \omega\frac{\partial C}{\partial x} + k_m(C_\infty - C) = 0, \tag{6-6}$$

where C = concentration of dissolved silica;
C_∞ = asymptotic concentration attained as $x \to \infty$ (can be, but is not necessarily, the saturation concentration—see Chapter 7)
$D(x)$ = mixing coefficient.

From the sediment-water interface ($x = 0$) to a depth of 3.5 centimeters the sediment is brown and indicates constant replenishment, by mixing, with dissolved oxygen. Below this depth the sediment is dark colored and anoxic. From these observations, Vanderborght et al. adopted a two-layer model. From $x = 0$ to $x = 3.5$ cm (layer 1) the value of $D(x)$ was set equal to a constant, D_{wc}, representing wave and current mixing. Below 3.5 cm (layer 2), $D(x)$ was assumed equal to the coefficient for molecular diffusion, D_s. Solution of (6-6) for each layer was done using the boundary conditions:

$x = 0$; $C = C_0$ (overlying water concentration),

$x = 3.5$ cm; $D_{wc}\left(\frac{\partial C}{\partial x}\right)_{layer\ 1} = D_s\left(\frac{\partial C}{\partial x}\right)_{layer\ 2}$

$x \to \infty$; $C \to C_\infty$.

The resulting analytical expressions for each layer were fit to the data of Figure 6-4. Using:

$$D_s = 10^{-6}\ \text{cm}^2\ \text{sec}^{-1},$$
$$\omega = 0.03\ \text{cm yr}^{-1},$$
$$C_\infty = 0.4\ \mu\text{mole cm}^{-3},$$

the best fit was obtained from the values:

$$k_m = 5 \times 10^{-7}\ \text{sec}^{-1},$$
$$D_{wc} = 10^{-4}\ \text{cm}^2\ \text{sec}^{-1}.$$

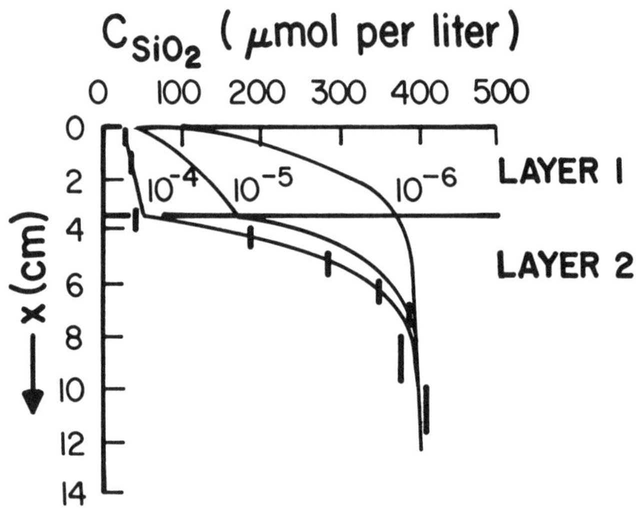

FIGURE 6-4. Fit of theoretical curves, according to the two-layer wave-and-current-stirring diagenetic model of Vanderborght et al. (1977a), to measured silica concentrations for Belgian coastal sediments. Values listed by each curve are those chosen for the wave-and-current mixing coefficient (in $cm^2\ sec^{-1}$). Best fit is provided by $D_{wc} = 10^{-4}\ cm^2\ sec^{-1}$. (After Vanderborght et al., 1977a.)

An idea of the variation of predicted curves using different values for D_{wc} is shown in Figure 6-4. Thus, Vanderborght et al. conclude that in the upper 3.5 centimeters of their sediments, mixing by waves and currents occurs at a rate about 100 times faster than that expected for molecular diffusion. This is distinctly higher than that calculated by Aller for bioturbational mixing of similar near-shore sediments, and indicates the effectiveness of physical disturbance in these sediments.

From this mixing coefficient, Vanderborght et al. (1977b) were able to apply a similar model to the distribution of dissolved nitrate in the same sediments. Again a two-layer model was employed with nitrification (bacterial oxidation of ammonia by O_2 to form nitrate) occurring in the top layer and denitrification (bacterial reduction of nitrate to N_2 and N_2O) occurring in the bottom layer. In other words:

For layer 1: $(0 \leqq x \leqq 3.5\ \text{cm})$

$$D_{wc}\frac{\partial^2 C}{\partial x^2} - \omega\frac{\partial C}{\partial x} + R_{nit} = 0. \tag{6-7}$$

For layer 2: $(x \geqq 3.5\ \text{cm})$

$$D_s\frac{\partial^2 C}{\partial x^2} - \omega\frac{\partial C}{\partial x} - k_{denit}C = 0, \tag{6-8}$$

where C = concentration of dissolved nitrate;

R_{nit} = rate of nitrification assumed to be a constant within layer 1;

k_{denit} = (first order) rate constant for denitrification.

The rate law assumed for denitrification is reasonable because at the low concentrations of nitrate found in the pore waters the Michaelis-Menten equation should be approximated by first order kinetics, and because metabolizable organic matter is present in large excess. The constant rate of nitrification rests upon the assumption of a relatively constant and excess supply of NH_4^+ and O_2 in this zone. Solution of (6-7) and (6-8) with the boundary conditions:

$$x = 0, \qquad C = C_0 \text{ (overlying water)},$$

$$x = 3.5 \text{ cm}, \qquad D_{wc}\left(\frac{\partial C}{\partial x}\right)_{layer\ 1} = D_s\left(\frac{\partial C}{\partial x}\right)_{layer\ 2},$$

$$x \to \infty, \qquad C \to 0$$

resulted in complex expressions which were fitted to the data shown in Figure 6-5. The best fit was obtained for the values:

$$R_{nit} = 1.5 \times 10^{-6}\ \mu\text{mole cm}^{-3}\text{ sec}^{-1},$$
$$k_{denit} = 5 \times 10^{-6}\text{ sec}^{-1}\ (158\text{ yr}^{-1}),$$

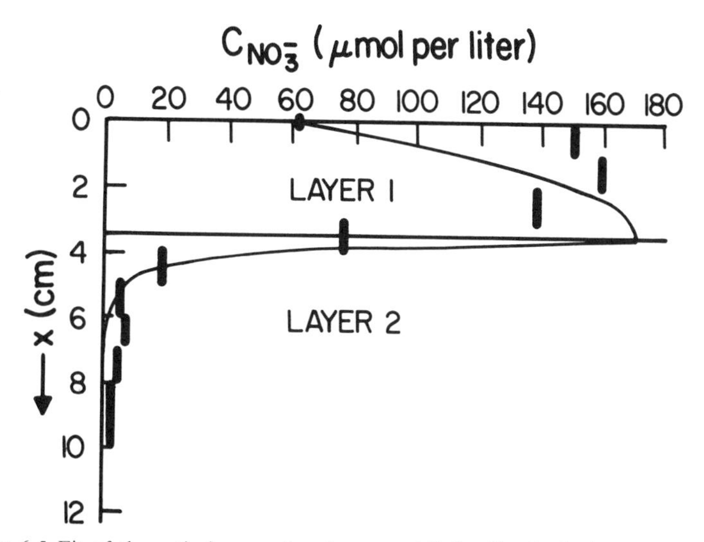

FIGURE 6-5 Fit of theoretical curve (two-layer model) for dissolved nitrate to sediment data for Belgian coastal sediments. (After Vanderborght et al., 1977b.)

using the values for ω, D_{wc}, and D_s obtained from the model for dissolved silica. The value for R_{nit} is in good agreement with actual measurements of nitrification ($R_{nit} = 1.8 \times 10^{-6}$ μmole cm^{-3} sec^{-1}) of similar sediments in the laboratory (Vanderborght and Billen, 1975). This provides a corroborative check on the model.

A two-layer model, similar to that used by Vanderborght et al., has been applied to the description of bacterial sulfate reduction and the sulfate concentration depth distribution in sediments from the FOAM site in Long Island Sound, U.S.A (Goldhaber et al., 1977). Here the upper zone was assumed to be mixed only by bioturbation and the lower zone to undergo only molecular diffusion. A complete diagenetic model for dissolved sulfate in the upper, bioturbated zone was not attempted because of seasonal changes in sulfate reduction rate and irrigation rate as discussed above for NH_4^+. However, a simple calculation permitted deduction of the *minimum* value of D_{BI} for sulfate. At the boundary ($x = 8$ cm) between the two zones the boundary condition used above applies:

$$D_{BI}\left(\frac{\partial C}{\partial x}\right)_{layer\ 1} = D_s\left(\frac{\partial C}{\partial x}\right)_{layer\ 2}. \tag{6-9}$$

For a typical summer profile, it was found that $(\partial C/\partial x)_{layer\ 1} < 0.07$ μmole cm^{-3} cm^{-1}. Combining this with the values $(\partial C/\partial x)_{layer\ 2} = 0.5$ μmole cm^{-3} cm^{-1} and $D_s = 3 \times 10^{-6}$ cm^2 sec^{-1}, one obtains:

$$D_{BI} > 2 \times 10^{-5}\ \text{cm}^2\ \text{sec}^{-1},$$

which is in reasonably good agreement with values of D_{BI} determined independently by Aller (1977; 1980) from modeling of ammonia fluxes out of the sediment.

Although detailed diagenetic modeling of dissolved *sulfate* in the zone of bioturbation of the FOAM sediments has not been attempted, a model has been proposed to describe *organic matter* decomposition accompanying sulfate reduction in this zone (Berner, 1980). The basic assumptions of the model are:

1. Organic matter decomposition accompanying sulfate reduction takes place via first order kinetics according to a two-G model (see the next section and Chapter 4), with one fraction of the organic matter (α) being decomposed within the zone of bioturbation and the other fraction (β) decomposed below the zone of bioturbation.
2. Bioturbation of metabolizable organic matter takes place according to a biodiffusion model with a fixed value of D_B.
3. Compactive water flow, porosity changes, etc. can be ignored, and SO_4^{--} does not undergo adsorption.

4. Steady state diagenesis can be assumed if mean annual concentrations of metabolizable organic matter and mean annual sulfate reduction rates are used.

With these assumptions, the appropriate diagenetic equation (from 3-73) for α-organic matter is:

$$D_B \frac{\partial^2 G_\alpha}{\partial x^2} - \omega \frac{\partial G_\alpha}{\partial x} - k_\alpha G_\alpha = 0, \tag{6-10}$$

where G_α = concentration (in mass per unit mass of total solids) of organic carbon which is metabolized within the zone of bioturbation;

D_B = biodiffusion coefficient for solids;

k_α = first order decay constant for α-organic matter.

Solution of (6-10) for the boundary conditions:

$$x = 0; \qquad G_\alpha = G_{\alpha 0},$$
$$x \to \infty; \qquad G_\alpha \to 0,$$

yields:

$$G_\alpha = G_{\alpha_0} \exp\left[\left(\frac{\omega - (\omega^2 + 4k_\alpha D_B)^{1/2}}{2D_B}\right) x\right]. \tag{6-11}$$

Now, according to the first order model, the rate of sulfate reduction accompanying the decomposition of α-organic matter is given by:

$$R_{SO_4} = -\mathscr{L} F k_\alpha G_\alpha, \tag{6-12}$$
$$R_{SO_{4_0}} = -\mathscr{L} F_0 k_\alpha G_{\alpha_0}, \tag{6-13}$$

where R_{SO_4} = rate of sulfate reduction in mass per unit volume of pore water per unit time;

$\mathscr{L}$ = stoichiometric coefficient relating the number of sulfate ions oxidized per atom of carbon oxidized ($\mathscr{L}$ normally equals 0.5—see next section);

$$F = \left(\frac{1 - \phi}{\phi}\right) \bar{\rho}_s.$$

(Since there is no compaction and steady state, $\phi = \phi_0$ and thus $F = F_0$.) Combining equation (6-11), (6-12), and (6-13):

$$R_{SO_4} = R_{SO_{4_0}} \exp\left[\left(\frac{\omega - (\omega^2 + 4k_\alpha D_B)^{1/2}}{2D_B}\right) x\right]. \tag{6-14}$$

Equation (6-14) was used to calculate the value of k_α for sediments in the bioturbation zone at the FOAM site. This was done (Berner, 1980) by comparing actual measurements of sulfate reduction rate as a function of depth, with those predicted by equation (6-14). The measured values were found to follow the empirical relation:

$$R_{SO_4} = R_{SO_{4_0}} \exp(-0.24x), \tag{6-15}$$

where R_{SO_4} here represents the mean annual rate of sulfate reduction at each depth. Identification of equation (6-15) with the theoretical expression (6-14) yields:

$$\frac{\omega - (\omega^2 + 4k_\alpha D_B)^{1/2}}{2D_B} = -0.24. \tag{6-16}$$

Substituting the average values for sediments of the FOAM site, $D_B = 5.5 \text{ cm}^2 \text{ yr}^{-1}$ and $\omega = 0.1 \text{ cm yr}^{-1}$, we can solve (6-16) for k_α:

$$k_\alpha = 0.60 \text{ yr}^{-1}. \tag{6-17}$$

This value is about five hundred times larger than the value (k_β) calculated for organic matter decomposition via sulfate reduction in sediments below the zone of bioturbation. In other words, organic matter undergoing decomposition within the zone of bioturbation is *much* more readily metabolized than that buried to greater depths. A reasonable explanation for this is that fresh, highly reactive organic matter newly deposited at the sediment-water interface (e.g., fallout from plankton blooms) is rapidly mixed downward into the sediment by bioturbation, and, as a result, high rates of sulfate reduction are maintained at shallow depths where bioturbation is active. At greater depths the fresh material is not present, because of a lack of downmixing, and consequently only less reactive residual organic compounds (β) are metabolized which results in much lower rates of sulfate reduction. Thus, the value of k for sulfate reduction decreases with depth, which is in general agreement with the predictions of Jorgensen (1978).

Diagenesis Below the Zone of Bioturbation

Considerably more diagenetic modeling has been applied to organic matter decomposition in sediments where there is negligible bioturbation. The reason for this is that both the chemistry and the processes of solute migration are far simpler. Fine-grained continental margin sediments, in general, undergo sufficiently rapid organic matter decomposition that any dissolved oxygen or nitrate introduced into the sediment solely by molecular diffusion (without bioturbation) is entirely consumed within a few millimeters of the sediment-water interface (Rhoads, 1974). Bioturbational irrigation is needed to maintain non-zero concentrations of O_2 and NO_3^- at measurable depths (normally a few centimeters). Thus, below the zone of bioturbation there is no mechanism for introducing O_2 or NO_3^- and, as a result, we are dealing usually with strictly anoxic microbial processes, in other words, with fermentation and sulfate reduction. Also, because of a lack of irrigation and particle mixing, we are dealing only with molecular diffusion and advection as the principal transport processes. These conditions apply not only to the deeper portions of anoxic sediments overlain by normal aerated seawater, but also to sediments deposited in suboxic to anoxic bottom water (<0.1 ml O_2 per liter) where, due to the lack of O_2, there is no benthic macrofauna, and thus no bioturbation (Rhoads and Morse, 1971).

Sulfate Reduction

Bacterial sulfate reduction is a common process of organic matter decomposition in continental margin sediments, both in and below the zone of bioturbation (e.g., Ostroumov et al., 1961; Kaplan et al., 1963). Seawater sulfate is used as an energy source by sulfate-reducing bacteria which reduce sulfate to H_2S via an overall process which can be schematically represented as:

$$2CH_2O + SO_4^{--} \rightarrow H_2S + 2HCO_3^-.$$

This process occurs only in the complete absence of oxygen; in other words the bacteria are obligate anaerobes. It has been shown in laboratory experiments (e.g., Goldhaber and Kaplan, 1974) that the sulfate-reducing bacteria themselves can only utilize small dissolved organic molecules. As a result, the overall reaction represented above must involve other fermentative microorganisms which break down the complex molecules of sedimentary organic matter ultimately to the simple molecules required by the sulfate reducers. Thus, when we refer to sulfate reduction in sediments we mean an overall process, which involves the activities of a whole community of interacting microorganisms, of which sulfate reducers constitute only a part.

Bacterial sulfate reduction in the absence of bioturbation has been modeled diagenetically by the author (Berner, 1964; 1971; 1974). The basic assumptions of the model are:

1. Organic matter decomposition accompanying bacterial sulfate reduction follows simple first order kinetics (the one-G model of Chapter 4) according to the reaction given above.
2. The only chemical reaction affecting dissolved sulfate in the pore water is bacterial reduction. In other words, there is no bacterial production, or mineral precipitation or dissolution. (The latter is reasonable for sediments of seawater salinity.)
3. Adsorption of sulfate is negligible.
4. Compaction, water flow, porosity gradients, etc. can be ignored.
5. Diffusion occurs via molecular processes only.
6. Steady state diagenesis is present.

Under these conditions, the appropriate diagenetic equations from (3-73), (3-74), and (6-12) are:

$$D_s \frac{\partial^2 C}{\partial x^2} - \omega \frac{\partial C}{\partial x} - \mathscr{L} F k G = 0 \quad \text{(sulfate)}, \tag{6-18}$$

$$-\omega \frac{\partial G}{\partial x} - kG = 0 \quad \text{(org. matter)}, \tag{6-19}$$

where C = concentration of dissolved sulfate;

G = concentration of metabolizable organic carbon (in moles per unit mass of total solids);

k = rate constant for sulfate reduction;

$\mathscr{L}$ = stoichiometric coefficient relating the number of moles of sulfate reduced per mole of organic carbon oxidized to CO_2 (normally $= \frac{1}{2}$);

$$F = \left(\frac{1-\phi}{\phi}\right)\bar{\rho}_s.$$

Both boundary conditions for the solution of equations (6-18) and (6-19) are a bit complicated and require some detailed discussion. The upper boundary is the base of the zone of bioturbation where, to simplify notation, we let $x = 0$. (This can be the sediment-water interface for sediments deposited in anoxic bottom waters.) Here the concentration of sulfate C_0 is equal to the concentration at the base of the zone of bioturbation which, due to irrigation, is often close to that for the overlying seawater. The concentration of metabolizable organic carbon here is denoted as G_0.

As we approach great depth one of two things may happen. Either metabolizable organic matter runs out and the concentration of sulfate approaches an asymptotic value so that $x \to \infty$, $G \to 0$, $C \to C_\infty$; or sulfate runs out. In the latter case it is possible to adopt the same lower boundary condition, but in this case the value of C_∞ becomes less than zero. In other words, for the purpose of curve fitting, one can visualize C_∞ as the asymptotic (negative) concentration of sulfate that would have been attained when $G \to 0$, had sulfate itself not run out. This approach is preferable to adopting a given depth $x = x'$, where $C = 0$, as a lower boundary condition because the depth x' is not controlled by external processes but rather by the internal processes of sulfate reduction. (This also leads to the necessity of choosing the particular, rather than the general, solution to equation (6-18).)

With these boundary conditions the appropriate solutions of (6-18) and (6-19) (for constant ϕ) are:

$$C = \left[\frac{\omega^2 F \mathscr{L} G_0}{\omega^2 + kD_s}\right] \exp\left[(-k/\omega)x\right] + C_\infty, \qquad (6\text{-}20)$$

$$G = G_0 \exp\left[-(k/\omega)x\right]. \qquad (6\text{-}21)$$

Also:

$$C_0 - C_\infty = \frac{\omega^2 F \mathscr{L} G_0}{\omega^2 + kD_s}. \qquad (6\text{-}22)$$

Note that the one-G model predicts an exponential decrease of sulfate concentration with depth.

Equation (6-20) has been fitted to sulfate concentration-vs-depth data for the non-bioturbated portions of many different anoxic sediments, ranging in sedimentation rate from 5 cm per year down to less than 0.001 cm per year (Berner, 1974; Toth and Lerman, 1977; Berner, 1978a). Values of G_0 calculated from the many curve-fits are all permissible in that they are lower than the measured values for total organic carbon at $x = 0$. Unfortunately, there is no independent way at present to determine G_0 directly, so that more exact checks of the model are impossible. However, curve fitting has resulted in values of (k/ω) which exhibit an interesting trend from sediment to sediment. Toth and Lerman (1977) have shown that (k/ω) correlates in a linear fashion with ω, or in other words:

$$k = A\omega^2, \qquad (6\text{-}23)$$

where A is an empirical constant ($= 0.04$ cm^{-2} yr). This is illustrated in Figure 6-6.

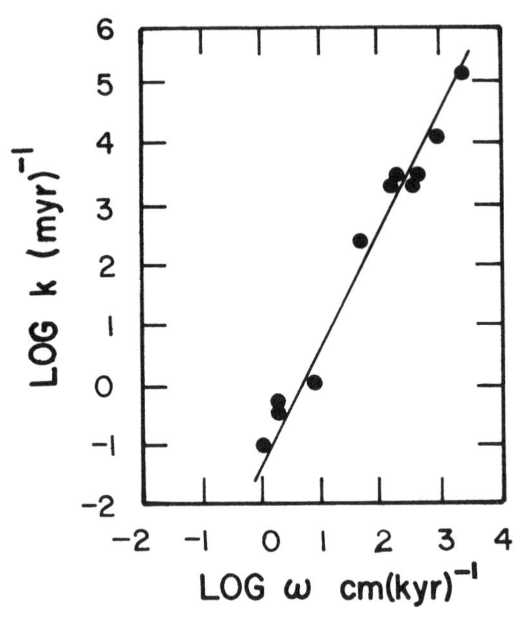

FIGURE 6-6. Log-log plot of k (rate constant for sulfate reduction) vs ω (rate of deposition) for marine sediments. Modified from Toth and Lerman (1977) to include additional data. (After Berner, 1978a)

Qualitative explanation of equation (6-23) can be given as follows. Higher sedimentation rates should result in more rapid burial and better preservation of readily metabolizable compounds which, at slower rates of deposition, would otherwise be destroyed by aerobic and anaerobic organisms living near the sediment-water interface within the zone of bioturbation. In this way rapid burial enables reactive organic substances to be furnished to sulfate-reducing bacteria plus associated fermentative microorganisms living below the bioturbation zone. Since k is a measure of organic matter reactivity, or metabolizability (see Chapter 4), one would therefore, expect a positive correlation between k and sedimentation rate.

Quantitative explanation of the empirical relation can be made in terms of the one-G diagenetic model. From (6-22) upon rearranging we obtain:

$$k = \left[\frac{\mathscr{L} F G_0}{D_s(C_0 - C_\infty)} - \frac{1}{D_s}\right]\omega^2. \tag{6-24}$$

The values of $\mathscr{L}$, F, and D_s are, within a factor of two, constant from one sediment to another. If the ratio $G_0/(C_0 - C_\infty)$ is also roughly constant (as will be shown below) we can explain relation (6-24) in terms of the one-G model and identify the expression within brackets in (6-24) with A of equation (6-23). The Toth-Lerman relationship thus lends further credence to the one-G model.

Besides the Toth-Lerman expression, an additional empirical relation has been reported by the author (Berner, 1978a). It was found that a very good linear correlation exists between the initial concentration gradient, just below the zone of bioturbation, $(\partial C/\partial x)_0$, and the rate of sedimentation. In other words:

$$\omega = -B\left(\frac{\partial C}{\partial x}\right)_0, \qquad (6\text{-}25)$$

where B is an empirical constant (for sediments with marine organic matter $B = 1.0\ \text{cm}^5\ \text{yr}^{-1}\ \mu\text{mole}^{-1}$). A plot of measured data is shown in Figure 6-7. Expression (6-25) is valid, whether or not the concentration-vs-depth profile is exponential. (Many profiles are linear and thus cannot be fitted uniquely by an exponential.)

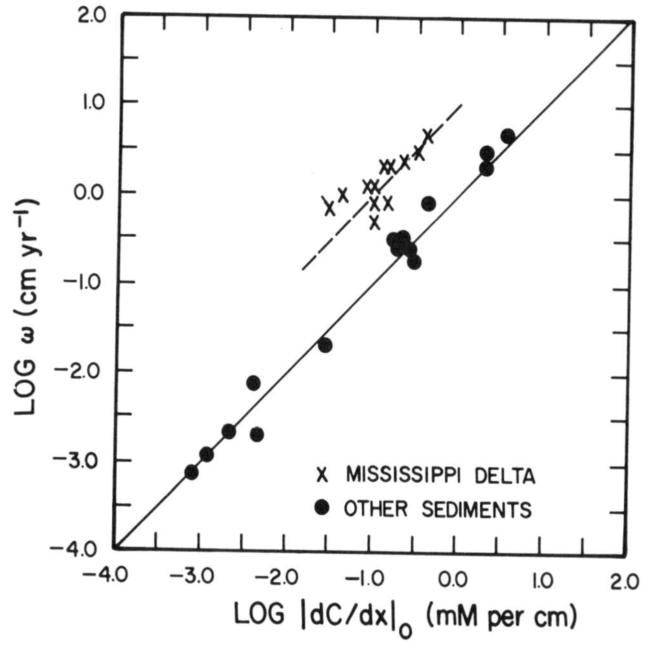

FIGURE 6-7. Log-log plot of the absolute value of initial sulfate gradient $|\partial C/\partial x|_0$ below the zone of bioturbation vs the rate of sedimentation ω. (After Berner, 1978a.)

Explanation of (6-25) has also been attempted in terms of the one-G model. To do this one must assume that the Toth-Lerman relation, equation (6-23), is valid for all sediments, even those that show straight-line profiles. In the latter case, it is assumed that the straight lines constitute the initial portions of exponentials. (Values of k obtained using measured values of ω and equation (6-23) in fact do give a permissible fit to almost all of the straight-line profiles.) The reasoning used is as follows. From equation (6-20):

$$\left(\frac{\partial C}{\partial x}\right)_0 = \frac{-k\omega F\mathscr{L}G_0}{\omega^2 + kD_s}. \tag{6-26}$$

Substituting (6-23) and rearranging:

$$\omega = -\left(\frac{1 + AD_s}{AF\mathscr{L}G_0}\right)\left(\frac{\partial C}{\partial x}\right)_0. \tag{6-27}$$

This is the same form as equation (6-25). Now, D_s, $\mathscr{L}$, and F are reasonably constant (within a factor of two) from one sediment to another and A is defined as being a constant. Then, if G_0 is essentially the same in all sediments, we can identify $(1 + AD_s/\mathscr{L}FG_0)$ with the empirical constant B so that:

$$G_0 = \frac{1 + AD_s}{F\mathscr{L}AB}. \tag{6-28}$$

From $A = 0.04\ \text{cm}^{-2}\ \text{yr}$, $B = 1.0\ \text{cm}^5\ \text{yr}^{-1}\ \mu\text{mole}^{-1}$, and the average values, $D_s = 100\ \text{cm}^2\ \text{yr}^{-1}$, $\mathscr{L} = 0.5$, and $F \approx 0.8\ \text{gm cm}^{-3}$, we obtain:

$$\begin{aligned} G_0 &= 310\ \mu\text{mole gm}^{-1} \\ &= 0.4\%\ C \text{ by dry weight.} \end{aligned}$$

This calculated value of G_0 is permissible in that for all sediments where $F \approx 0.8\ \text{gm cm}^{-3}$, the total organic carbon concentration at the base of the zone of bioturbation is greater than 0.4%. (For several deep-sea sediments showing sulfate reduction, total organic carbon contents are lower than 0.4% but values of G_0 are correspondingly less because of higher F values.) Thus, again the one-G model can be used to explain observed sulfate-vs-depth profiles.

The above agreements between theory and field measurements rest upon the assumption of constant G_0 from one sediment to another. At first sight this seems unreasonable since the sediments described in Figure 6-8 have total organic carbon contents varying by more than a factor of 20. Nevertheless, constant G_0 is not unreasonable. This can be seen from consideration of the general or multi-G model of Chapter 4:

$$R_{biol} = k_1G_1 + k_2G_2 + k_3G_3 + \cdots \tag{6-29}$$

Assume that only three groups of organic substances, G_1, G_2, and G_3, with $k_1 > k_2 > k_3$, account for all decomposable organic matter supplied to the sulfate-reducing community of microorganisms in all sediments. In rapidly deposited sediments, all three substances may be buried and bacteria will attack G_1 material first, because of its high k value (i.e., the term k_1G_1 dominates the right-hand side of equation 6-29). If sulfate is exhausted before the stock of G_1 is used up (in most sediments studied, sulfate actually drops to zero), then the rate of reduction will be expressed solely in terms of k_1G_1 and G_0 will refer only to G_{1_0} and *not* to the sum of $G_{1_0} + G_{2_0} + G_{3_0}$. In this case the leftover G_1 plus all G_2 and G_3 are decomposed by methanogens living below the zone of sulfate reduction (see below). By contrast, in sediments deposited very slowly all G_1 and G_2 may be consumed prior to burial of organic matter into the zone of sulfate reduction. Then, only G_3 is available and the rate of reduction, expressed as k_3G_3 will be lower and G_0 will refer only to G_{3_0}. At intermediate sedimentation rates G_2 and G_3 may be available but sulfate runs out before all G_2 is consumed. Here G_0 would refer to G_{2_0}. Note that in all cases the value calculated for G_0 would refer only to *each subgroup* G_1, G_2, or G_3 and not to their sum. If these subgroups have about the same concentration, then the value of G_0 calculated from multi-G type modeling would be about the same for all sediments, as is the case for one-G modeling, discussed above. In this way (using many subgroups) the constancy of G_0 from one sediment to another may be explained.

Murray et al. (1978) have recently applied the one-G type model to the study of sulfate reduction in sediments of Saanich Inlet, British Columbia. This study is especially noteworthy because of the inclusion of compaction in the diagenetic equation. The sediments studied show considerable porosity reduction in the top 10 centimeters (see Figure 6-8) where most sulfate reduction also occurs. (A drop of ϕ from 0.96 to 0.92 appears small but the change in the important parameter $(1 - \phi)$ is a *factor of two*.) Also, the presence of restricted circulation with periodic anoxia inhibits the development of a bottom-burrowing fauna and, as a result, bioturbation can be ignored. Besides neglecting bioturbation, Murray et al. also assumed that steady state was attained with respect to both sulfate and porosity and that molecular diffusion followed the relation (see equation 3-51):

$$D_s = D_0\phi^2, \tag{6-30}$$

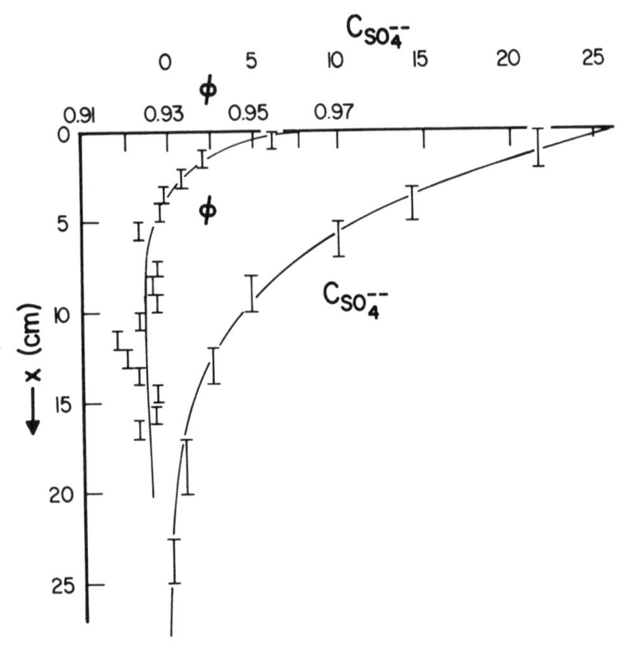

FIGURE 6-8. Porosity and dissolved sulfate data, fit by theoretical curves, for sediments of Saanich Inlet (British Columbia). See text for details. Sulfate in mmoles per liter. (Adapted from Murray et al., 1978.)

where D_0 = diffusion coefficient for pure pore water. Under these conditions the appropriate diagenetic equations, from (3-20), (3-73), (3-74), (6-12), and (6-30), are:

$$\frac{\partial\left(\phi^3 D_0 \frac{\partial C}{\partial x}\right)}{\partial x} - \phi_x \omega_x \frac{\partial C}{\partial x} - \phi \mathscr{L} F k G = 0, \tag{6-31}$$

$$-\omega \frac{\partial G}{\partial x} - kG = 0. \tag{6-32}$$

To solve (6-31) and (6-32), Murray et al. fit the data of Figure 6-8 to the empirical expression:

$$\phi = \phi_x + (\phi_0 - \phi_x) \exp(-0.55x), \tag{6-33}$$

where x is in centimeters. Also, for steady state compaction:

$$\omega = \left(\frac{1 - \phi_x}{1 - \phi}\right) \omega_x. \tag{3-19}$$

Combining equations (6-33) and (3-19) with (6-31) and (6-32), the final expressions used are:

$$\frac{\partial\left\{[\phi_x - (\phi_0 - \phi_x)\exp(-0.55x)]D_0 \dfrac{\partial C}{\partial x}\right\}}{\partial x} - \phi_x\omega_x \frac{\partial C}{\partial x}$$

$$- [1 - \phi_x - (\phi_0 - \phi_x)\exp(-0.55x)]\bar{\rho}_s\mathscr{L}kG = 0, \quad (6\text{-}34)$$

$$\left[\frac{(1 - \phi_x)\omega_x}{1 - \phi_x - (\phi_0 - \phi_x)\exp(-0.55x)}\right]\frac{\partial G}{\partial t} - kG = 0. \quad (6\text{-}35)$$

These equations were solved numerically using the values: $\phi_0 = 0.963$, $\phi_x = 0.924$, $D_0 = 63\ \text{cm}^2\ \text{yr}^{-1}$, $\omega_x = 4.04\ \text{cm yr}^{-1}$, $\bar{\rho}_s = 2.5$, and $\mathscr{L} = 0.5$, and the boundary values $C_0 = 24.8\ \mu\text{mole cm}^{-3}$ and $C_\infty = 0$. Best fits to the measured sulfate concentration data shown in Figure 6-8 was provided by the values:

$$k = 0.189\ \text{yr}^{-1},$$
$$G_0 = 2900\ \mu\text{mole gm}^{-1}\ (3.5\%).$$

This calculated G_0 value is too high since the drop in total carbon from $x = 0$ to $x = 20$ cm is only about 0.9%. Murray et al. explain this discrepancy in terms of additional organic carbon supplied to the sulfate-reducing community by diffusion of methane from below. This is possible in that high concentrations of dissolved methane are found in these sediments just below the zone of sulfate reduction. If methane is truly a major source of G, then the assumption of non-diffusability of G, explicit in the one-G model, is incorrect and equation (6-31) above must include an additional term for the diffusion of G.

The results of Murray et al., however, may also be explained in terms of non-steady state diagenesis without invoking a diffusable source of G. For example, there may have been unusually rapid deposition of sediment containing higher-k type organic matter over the top 5 centimeters. (In other words, a step-like increase in G_0 and k may have occurred relatively recently.) This idea is supported by the observation of relatively constant ^{210}Pb concentration over the top 5 centimeters as compared to greater depths, which suggests very rapid sedimentation over this depth zone. If true, the sharp decrease of SO_4 concentration might represent a relatively recent phenomenon and, thus, not require a large integrated amount of sulfate reduction (and consequent high G_0) over the top 5–10 centimeters, as would be the case for steady state diagenesis. Unfortunately, the sulfate curve itself gives no clue as to any possible non-steady state changes.

For *periodic* variations (not step functions as above) Lasaga and Holland (1976) have presented a method for deducing whether or not changes in

G_0 can be seen in terms of fluctuations (slope reversals) in the profiles of dissolved sulfate (and other species) vs depth. Using a one-G type, non-steady state model without bioturbation, they assumed that at $x = 0$:

$$G_0 = \bar{G}_0 + L \sin 2\pi\alpha t, \tag{6-36}$$

where $\bar{G}_0$ = average long-term value of G_0;
α = frequency of G_0 fluctuation;
L = amplitude of G_0 fluctuation.

From a simple argument they were able to show that in order to see non-steady state fluctuations in the curve of sulfate-vs-depth resulting from fluctuations in G, the following inequality must be met:

$$\alpha < \frac{\omega^2}{D_s}. \tag{6-37}$$

Otherwise, the curve will be indistinguishable from that for steady state at constant $\bar{G}_0$. A table for different values of ω has been constructed (Table 6-1) to illustrate this inequality. Note that for normal ω values, annual (seasonal) variations in G_0, where $\alpha = 1\ yr^{-1}$, are far too rapid to be detected as reversals in the curve of sulfate vs depth. Diffusion over the distances representing annual layers is too fast to allow preservation of concentration differences resulting from different reduction rates. Also, many longer term variations are dampened out by diffusion. For a typical near-shore sediment, deposited at a rate of $\omega = 0.2$ cm yr^{-1}, the highest frequency of G_0 variation detectable as reversals in the sulfate profile would be 4×10^{-4} per year. This represents a time of 1,250 years between maxima and minima which is equivalent to 250 centimeters of deposited sediment.

TABLE 6-1

Values of α necessary to see fluctuations in G_0 in terms of reversals in the concentration-vs-depth profiles of dissolved sulfate. Calculations are for $D_s = 100\ cm^2\ yr^{-1}$.

ω (cm yr^{-1})	α (yr^{-1})
0.0001	10^{-10}
0.001	10^{-8}
0.01	10^{-6}
0.1	10^{-4}
1.0	10^{-2}

The Lasaga and Holland argument applies not only to sulfate reduction but to all other processes, such as ammonia and phosphate liberation, associated with organic matter decomposition. However, it is only valid in the absence of bioturbation. With bioturbation, short-term fluctuations in G_0 (and especially k) may bring about fluctuating depth profiles of sulfate, ammonia, and so on, because of rapid down-mixing of fresh high-k organic material over a depth range much greater than that representing one year of sediment accumulation (e.g., see Aller, 1977).

One approach to bacterial sulfate reduction below the zone of bioturbation which avoids any assumptions as to rate laws is that adopted by Goldhaber et al. (1977) for sediments of the FOAM site in Long Island Sound, U.S.A. Here rates of sulfate reduction R_{SO_4} are calculated at each depth from the steady state diagenetic equation for sulfate, which (at constant porosity and constant diffusion coefficient) is:

$$D_s \frac{\partial^2 C}{\partial x^2} - \omega \frac{\partial C}{\partial x} = R_{SO_4}. \qquad (6\text{-}38)$$

To accomplish this, measured C vs x data were fitted to a fifth order polynomial by computer and the values of $\partial C/\partial x$ and $\partial^2 C/\partial x^2$ calculated for each depth. The data could be fitted almost as well by an exponential, but this was avoided since an assumption of exponential decrease with depth, by necessity, implies a one-G type model (Berner, 1979). Also, D_S was measured in situ in the sediments under study and ω was determined by stratigraphic correlation and radiometric dating.

The results are shown in Table 6-2 where they are compared to actual measurements (good to a factor of two) of sulfate reduction rate made on the same sediments. As can be seen, considering that Goldhaber et al.

TABLE 6-2

Rates of sulfate reduction, measured and calculated via equation (6-38), for sediments of the FOAM site, Long Island Sound, U.S.A. Values used for calculation: ω = 0.3 cm yr^{-1}, $D_s = 4 \times 10^{-6}$ cm^2 sec^{-1}. (Data from Goldhaber et al., 1977.)

	mmol per liter per yr	
Depth (cm)	*Calculated rate*	*Measured rate*
10	2.2	—
15	—	1.8 ± 0.3
20	1.5	—
25	—	1.8 ± 0.6
30	1.0	—

assumed steady state for a sediment (FOAM site) which shows an obvious non-steady state distribution of total organic matter, the agreement between measured and calculated rate is amazingly good. It should be noted that the calculated value of R_{SO_4} is very insensitive to the value chosen for ω, which is fortunate since there is some dispute over sedimentation rates in this portion of Long Island Sound (Benninger et al., 1979).

An important consequence of bacterial sulfate reduction heretofore mentioned only briefly is the formation of pyrite, FeS_2. Some of the hydrogen sulfide formed by the microorganisms reacts with detrital iron minerals, via a series of steps (e.g., Goldhaber and Kaplan, 1974), to form pyrite. Some H_2S also escapes upward via diffusion and is eventually destroyed by oxidation near the sediment-water interface. If, however, sufficient reactive iron is present to trap essentially all H_2S formed, then little escapes, and the amount of pyrite can thereby be calculated by means of diagenetic modeling. Complete trapping is best accomplished in the absence of bioturbation, which otherwise aids in H_2S loss and oxidation. Thus, modeling of pyrite is best done below the zone of bioturbation.

The total amount of sulfate reduced $\sum S$ during burial of a layer from $x = 0$ to its present depth, for steady state diagenesis, is given by the expression:

$$\sum S = \int_0^x \frac{R_{SO_4}}{\omega} dx. \tag{6-39}$$

If all sulfide formed by sulfate reduction is trapped as pyrite sulfur, then the pyrite concentration in a given layer is simply equal to $\sum S$. Evaluation of R_{SO_4} can be done either by the method of Goldhaber et al., described above, or through the use of the steady-state one-G model. According to the latter:

$$\begin{aligned} R_{SO_4} &= \mathscr{L}FkG, \\ &= \mathscr{L}FkG_0 \exp\left(\frac{-k}{\omega}x\right). \end{aligned} \tag{6-40}$$

Insertion of (6-40) into (6-39), and integration for constant ω (e.g., no compaction) yields:

$$\sum S = \mathscr{L}FG_0\left[1 - \exp\left(\frac{-k}{\omega}x\right)\right]. \tag{6-41}$$

This equation indicates that the concentration of pyrite should increase with depth according to the same exponential used to describe sulfate reduction. In other words, the curves of sulfate and pyrite vs depth should be mirror images of one another. In general, most pyrite is formed within the zone of bioturbation; and as a result, little reactive iron is available for further pyrite formation at greater depths, and complete trapping of

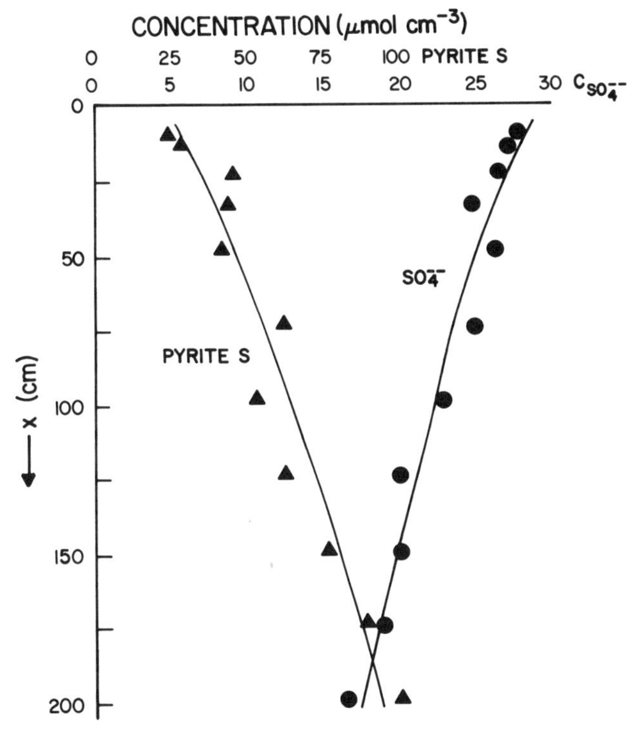

FIGURE 6-9. Plots of pyrite sulfur and dissolved sulfate concentration vs depth for Gulf of Mexico sediments. Both curves fitted with exponentials with the same value of $k/\omega = 0.0025\ \text{cm}^{-1}$. (Adapted from Filipek and Owen, 1980.)

H_2S does not occur. However, in some sediments mirror-image-type sulfate and pyrite curves are found, indicating major pyrite formation at depth. An example is shown in Figure 6-9 taken from the work of Filipek and Owen (1980). G_0 calculated from (6-41) and G_0 calculated from (6-20), using the sulfate and pyrite data of Figure 6-9 plus values of D_s and ω, are in good agreement for this sediment.

Ammonia Formation

Ammonia is formed below the zone of bioturbation in continental margin sediments by the decomposition of organic nitrogen compounds. Here, because dissolved oxygen is normally lacking, ammonia does not undergo oxidation to NO_2^- and NO_3^- (as it does in the bioturbation zone) and, as a result, it can build up to high concentrations. Since most of the sediments are in the pH range 7–8, ammonia is present as NH_4^+, the predominant ion under these conditions (e.g., see Berner, 1971). Being a cation, NH_4^+ undergoes exchange reactions with other cations associated with the sediment particles. Thus, in contrast to sulfate, adsorption (ion exchange) must be considered when attempting to model the diagenetic distribution of NH_4^+. However, like sulfate, NH_4^+ is not appreciably involved in authigenic mineral formation. The author (Berner, 1971; 1974) has modeled the diagenetic depth distribution of NH_4^+ in marine sediment pore waters, in the absence of bioturbation, according to a one-G organic decomposition model similar to that used for sulfate but including the effects of adsorption. The assumptions are:

1. Decomposition of organic nitrogen to ammonia occurs via first order kinetics according to the nitrogen analogue of the one-G model. In other words:

$$\frac{dN}{dt_{biol}} = -kN, \tag{6-42}$$

   where N refers to the concentrations of non-diffusable, metabolizable organic nitrogen (e.g., in proteins) in mass of N per unit mass of total solids, and k is the first order decomposition rate constant. Material denoted as N is assumed to be converted entirely to ammonia without appreciable build-up of intermediate dissolved nitrogen compounds (e.g., amino acids). Thus, the rate of organic nitrogen decomposition is set equal to the rate of ammonia formation.
2. Equilibrium adsorption is present and follows a simple linear isotherm.
3. Precipitation of ammonium ion in authigenic minerals does not occur (reasonable for almost all circumstances).
4. The sediment under study is entirely anoxic so that oxidation of NH_4^+ to NO_3^- and NO_2^- does not take place.
5. Compaction, water flow, porosity gradients, etc. can be ignored.
6. Diffusion occurs via molecular processes only.
7. Steady state diagenesis is present.

Under these conditions the proper diagenetic equations from (3-73) and (4-53) are:

$$-\omega\frac{\partial N}{\partial x} - kN = 0, \tag{6-43}$$

$$\left(\frac{D_s}{1+K_N}\right)\frac{\partial^2 C}{\partial x^2} - \omega\frac{\partial C}{\partial x} + \frac{FkN}{1+K_N} = 0, \tag{6-44}$$

where C = concentration of dissolved NH_4^+;

K_N = equilibrium adsorption constant for NH_4^+ ion ($=FK'_N$). K_N is expressed in mass adsorbed per unit volume of pore solution.

Boundary conditions for the solution of equations (6-43) and (6-44) are analogous to those for sulfate. At $x = 0$, $N = N_0$ and concentration C_0 is set equal to the value at the base of the zone of bioturbation. At depth, metabolizable organic nitrogen becomes consumed and:

$$x \to \infty, \quad N \to 0, \quad C \to C_\infty.$$

With these boundary conditions, the solutions of equations (6-43) and (6-44) are:

$$N = N_0 \exp[(-k/\omega)x], \tag{6-45}$$

$$C = \frac{\omega^2 F N_0}{D_s k + (1 + K_N)\omega^2}\{1 - \exp[(-k/\omega)x]\} + C_0. \tag{6-46}$$

Also:

$$C_\infty - C_0 = \frac{\omega^2 F N_0}{D_s k + (1 + K_N)\omega^2}. \tag{6-47}$$

Rosenfeld (1980) has presented extensive data for sediments of Long Island Sound, U.S.A, which can be used to check equations (6-45) and (6-46). For sediments of the Sachem Harbor site (where bioturbation is minimal) Rosenfeld has fitted to measured concentrations of dissolved NH_4^+ (Figure 6-10), the empirical expression:

$$C = 8.4[1 - \exp(-0.016x)] + 1.5, \tag{6-48}$$

where C is in unit of μmole per cm^3 and x is measured from the base of the zone of bioturbation. Identifying (6-48) with the theoretical expression (6-46) we obtain:

$$\frac{k}{\omega} = 0.016 \text{ cm}^{-1}, \tag{6-49}$$

$$\frac{\omega^2 F N_0}{D_s k + (1 + K_N)\omega^2} = 8.4 \ \mu\text{mol cm}^{-3}. \tag{6-50}$$

Rosenfeld conducted adsorption experiments with the same sediments and demonstrated rapid reversibility of NH_4^+ adsorption with $K_N = 1.3$. He also estimated D_s for NH_4^+ from direct measurements of D_s for SO_4^{--} in nearby sediments and use of the ratio $D_s(NH_4^+)/D_s(SO_4^{--})$ in sea-water (Li and Gregory, 1974). The value of ω was found to be between the limits 0.5 and 1.0 cm per yr using palynological measurements. We will use the value 0.8 cm yr^{-1} here. From porosity measurement for this sediment $F = 0.83$ gm cm^{-3}. Combining these data with equations (6-49) and (6-50) we obtain:

$$k = 0.013 \text{ yr}^{-1}$$
$$N_0 = 62.3\ \mu\text{mol gm}^{-1} = 0.087\%.$$

From these data the predicted expression for metabolizable organic nitrogen vs depth, using equation (6-45) is:

$$N = 62.3 \exp(-0.016x). \qquad (6\text{-}51)$$

Rosenfeld also measured total organic nitrogen vs depth in the same sediment and results are also shown in Figure 6-10. This data can thus be used to check the validity of the model calculation above by assuming

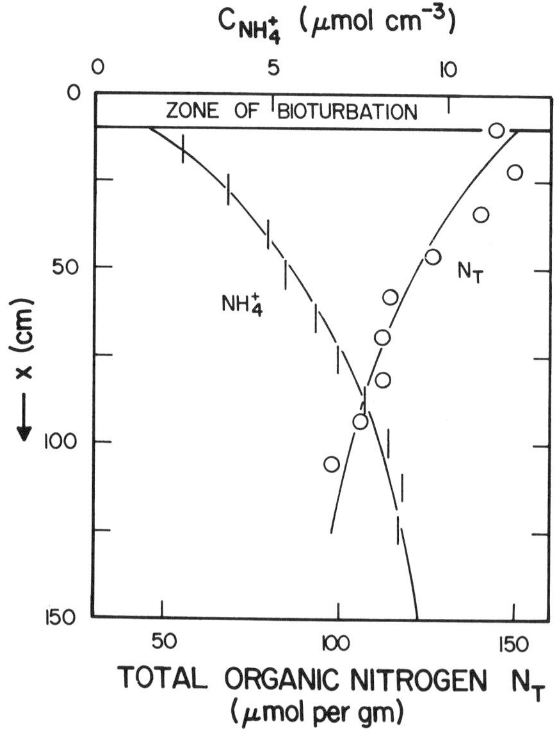

FIGURE 6-10. Measured data for dissolved NH_4^+ and total organic nitrogen fitted by theoretical curves according to the diagenetic model discussed in the text. The good fit to the data by both curves provides a positive check on the model. (Data from Rosenfeld, 1980 for the SACHEM site of Long Island Sound.)

that the asymptotic value of organic nitrogen represents non-metabolizable material. From Figure 6-10 this amounts to about 88 μmol gm^{-1}. Adding this to equation (6-51) we obtain the theoretical expression for *total* organic nitrogen vs depth:

$$N_T = 88 + 62.3 \exp(-0.016x). \quad (6\text{-}52)$$

This curve is plotted on Figure 6-10 and, as can be seen, fits the measured data well below the top few centimeters. (In this zone problems due to seasonal fluctuations and/or bioturbation may be present). This good fit of the theoretical curve provides independent evidence that the simple steady state one-G model is a good description of anoxic ammonia production in this sediment.

Rosenfeld also estimated, independently, the rate of total (dissolved + adsorbed) ammonia production from measurements of ammonia build-up in natural sediment samples from Sachem Harbor incubated in the laboratory. His values are compared in Table 6-3 with those calculated from the one-G theoretical expression:

$$\frac{dC}{dt_{biol}} = -FkN, \quad (6\text{-}53)$$

which from (6-45) can be written:

$$\frac{dC}{dt_{biol}} = -FkN_0 \exp\left(\frac{-k}{\omega}x\right). \quad (6\text{-}54)$$

Table 6-3 shows good agreement between measured production rates and those predicted from the theoretical model. This agreement gives further credence to the validity of the model.

TABLE 6-3

Rates of ammonia production measured in the laboratory and calculated via equation (6-54) for sediments from the SACHEM site, Long Island Sound. Range in calculated rates reflects range in estimates of ω. (Data from Rosenfeld, 1980.)

	mmol per liter per yr	
Depth (cm)	*Calculated rate*	*Measured rate*
15	0.52–0.66	0.65
25	0.44–0.57	0.48

To this point ammonia production has been treated independently of other bacterial processes occurring simultaneously in the sediment. This approach is appropriate if interest is directed solely to nitrogen. However, decomposable nitrogen is contained in organic carbon compounds, and the destruction of these compounds is what is really involved in the production of ammonia. Under the anoxic conditions found below the zone of bioturbation in continental margin sediments the two major overall processes of organic matter decomposition are sulfate reduction and methane formation (see Chapter 4). These two processes take place in succession, with major methane production occurring at depths below which sulfate concentrations drop to zero.

Since ammonia production is tied up with organic matter decomposition, the question arises as to the stoichiometric relation between carbon and nitrogen decomposition during either sulfate reduction or methane formation. In other words, does the material represented as G have a constant or variable $C:N$ ratio during decomposition. Knowledge of the $C:N$ ratio provides additional clues as to the nature of the metabolizable material. (Note that the $C:N$ ratio here refers only to the metabolizable fraction and is, in general, different from that measured for total organic matter—e.g., see Rosenfeld, 1980.) Also it is of interest to inquire whether the rate constant for ammonia release, as well as the $C:N$ ratio, changes when the overall decomposition process changes (downward) from sulfate reduction to methane production.

Within the zone of sulfate reduction the $C:N$ ratio of organic matter undergoing decomposition, by sulfate-reducing bacteria plus associated fermentative microorganisms, can be deduced from dissolved sulfate and ammonia data by means of diagenetic modeling (Berner, 1977). This is done through the use of the differential form of equations (6-20) and (6-46):

$$dC_S = \left[\frac{-k_S\omega F\mathscr{L}G_0}{\omega^2 + k_S D_{sS}}\right]\exp\left[(-k_S/\omega)x\right]dx, \tag{6-55}$$

$$dC_N = \left[\frac{k_N\omega F N_0}{\omega^2(1 + K_N) + k_N D_{sN}}\right]\exp\left[(-k_N/\omega)x\right]dx, \tag{6-56}$$

where D_{sS} = diffusion coefficient for SO_4^{--};
D_{sN} = diffusion coefficient for NH_4^+;

and subscripts S and N refer to sulfate and ammonium respectively. Combining these two expressions:

$$\frac{dC_S}{dC_N} = \frac{-G_0\mathscr{L}k_S}{N_0 k_N}\left[\frac{\omega^2(1 + K_N) + k_N D_{sN}}{\omega^2 + k_S D_{sS}}\right]\left\{\frac{\exp\left[(-k_S/\omega)x\right]}{\exp\left[(-k_N/\omega)x\right]}\right\}. \tag{6-57}$$

Also, from (6-21) and (6-45):

$$\frac{G_0}{N_0} = \frac{G \exp[(-k_N/\omega)x]}{N \exp[(-k_S/\omega)x]}. \tag{6-58}$$

so that upon substitution in (6-57) and solving for G/N (the $C:N$ ratio of decomposing organic matter):

$$\frac{G}{N} = \frac{-k_N}{\mathscr{L} k_S}\left[\frac{\omega^2 + k_S D_{sS}}{\omega^2(1 + K_N) + k_N D_{sN}}\right]\frac{dC_S}{dC_N}. \tag{6-59}$$

This expression thus allows calculation of G/N at each depth as a function of dC_S/dC_N.

Plots of dissolved sulfate vs dissolved ammonia are often linear for sediment pore waters (e.g., see Martens et al., 1978; Hartmann et al., 1973; Suess, 1976). This means that in equation (6-59), dC_S/dC_N is a constant, independent of depth. If this is true, then G/N must be constant with depth since all other parameters in (6-59) are also constant (by the model assumptions). Constancy of dC_S/dC_N, in equation (6-57), means that no depth functionality is present and that the exponential terms must cancel. This can happen only if $k_S = k_N$. Thus, for sediments showing straight-line plots of sulfate vs ammonia:

$$\frac{G}{N} = \frac{-1}{\mathscr{L}}\left[\frac{\omega^2 + kD_{sS}}{\omega^2(1 + K_N) + kD_{sN}}\right]\frac{dC_S}{dC_N}, \tag{6-60}$$

where $k = k_S = k_N$. This equation shows how the $C:N$ ratio of organic matter decomposed during sulfate reduction can be deduced.

Rosenfeld (1980) has shown that plots of C_S vs C_N are linear within the zone of sulfate reduction in sediments from the Sachem Harbor site. Substituting the slope of this plot along with values of k, ω, D_{sS}, D_{sN}, and K_N into (6-60) and using the conventional value of $\mathscr{L} = \frac{1}{2}$ (2 moles of carbon oxidized for each mole of sulfate reduced), Rosenfeld obtained for one core:

$$\frac{G}{N} = 6.7.$$

This value agrees with results from laboratory decomposition experiments using the same sediments. It is also distinctly lower than the ratio of measured total organic carbon to total organic nitrogen (15) and shows that organic compounds enriched in nitrogen are preferentially decomposed during sulfate reduction at the Sachem Harbor site. Since sulfate reduction rates are high here relative to most other continental margin sediments, it suggests that higher values of k, or greater metabolizability, may be associated with organic compounds enriched in nitrogen.

Below the depth where sulfate disappears ammonia concentrations continue to build up. This is shown by the data of Berner et al. (1970), Rosenfeld (1980), and Murray et al. (1978). Apparently the breakdown of organic compounds by fermentative microorganisms accompanying sulfate reduction continues into the zone of methane production. In the sediments (from the New England coastal region) studied by Berner et al. and Rosenfeld, ammonia production rate did not change when passing downward from the sulfate into the methane zone. This suggests that the fermentative organisms that decompose high molecular weight compounds to the low molecular weight molecules used by sulfate reducers (Goldhaber and Kaplan, 1974) are similar in both zones and that the $C:N$ ratio is also similar. Unfortunately, because of methane saturation and bubble formation this suggestion cannot be easily tested using ordinary diagenetic modeling for these sediments.

Murray et al. (1978), in contrast to the above results, note an inflection in the dissolved ammonia-vs-depth curve at about the same depth where sulfate disappears in sediments from Saanich Inlet, British Columbia. Their data indicate a definite change in the rate of production of ammonia between the sulfate and methane zone. Thus, no generalization can be made at present about the difference, or lack of difference, in $C:N$ ratio and ammonia production rate between the zones of sulfate reduction and methane production.

Phosphate Diagenesis

In many ways phosphate diagenesis in continental margin sediments parallels that of ammonia. Phosphate tied up in organic compounds is liberated to solution during decomposition by bacteria. It also undergoes appreciable adsorption on the surfaces of particles. However, phosphate differs from ammonia in that it commonly precipitates to form various authigenic minerals, chiefly apatite ($Ca_5(PO_4)_3(OH,F)$) and vivianite $Fe_3(PO_4)_2 \cdot 8H_2O$, and under oxic conditions it does not undergo oxidation as does ammonia. Also, under oxic conditions, phosphate is much more strongly adsorbed (on ferric oxides) whereas NH_4^+ is not. Upon removal of oxygen and reduction of iron, the phosphate is liberated to solution. Below the zone of bioturbation, oxic conditions are rare in muds, and here the major processes controlling phosphate concentration are diffusion, deposition, adsorption, organic matter decomposition, and authigenic mineral precipitation. The one-G diagenetic model, discussed above for sulfate and ammonia, has also been applied to the dissolved phosphate concentration distribution in sediments (Berner, 1974). The basic assumptions of the model are:

1. Decomposition of organic phosphorus to dissolved phosphate occurs via first order kinetics according to the direct analogue of the model used for ammonia production from organic nitrogen (see above).
2. Equilibrium adsorption is present and follows a simple linear isotherm.
3. Precipitation of phosphate to form authigenic minerals takes place via the simple linear rate law: $R_{pptn} = k_m(C - C_{eq})$.
4. The sediments under study are entirely anoxic so that strong adsorption by ferric oxides does not occur.
5. Compaction, water flow, porosity gradients etc. can be ignored.
6. Diffusion occurs via molecular processes only.
7. Steady state diagenesis is present.

Under these conditions the appropriate diagenetic equations from (3-73) and (4-53) are:

$$-\omega \frac{\partial P}{\partial x} - kP = 0, \tag{6-61}$$

$$\left(\frac{D_s}{1 + K_P}\right)\frac{\partial^2 C}{\partial x^2} - \omega \frac{\partial C}{\partial x} + \frac{FkP}{1 + K_P} - \frac{k_m(C - C_{eq})}{1 + K_P} = 0, \tag{6-62}$$

where C = concentration of dissolved phosphate;

C_{eq} = concentration at saturation with the authigenic precipitate;

P = concentration of metabolizable, solid (non-diffusible) organic phosphorus in mass per unit mass of total sediment solids;

k = rate constant for organic phosphorus decomposition;

k_m = authigenic mineral precipitation rate constant;

K_P = equilibrium adsorption constant for phosphate.

Because of the precipitation of authigenic minerals, the lower boundary condition for solution of these equations is somewhat different from that for the case of ammonia. The boundary conditions are:

$$x = 0; \qquad P = P_0; \qquad C = C_0,$$

$$x \to \infty; \qquad P \to 0; \qquad C \to C_{eq}.$$

Note that at depth the concentration approaches that for saturation with respect to the authigenic phase, which is lower than the value that would be attained (C_∞) in the absence of precipitation. Solutions of (6-61) and

(6-62) for these boundary conditions are:

$$P = P_0 \exp[(-k/\omega)x], \tag{6-63}$$

$$C = \left[\left(\frac{FkP_0\omega^2}{D_s k^2 + (1 + K_P)\omega^2 k - k_m\omega^2}\right) - (C_{eq} - C_0)\right] \times \exp\left\{\frac{(1 + K_P)\omega - [(1 + K_P)^2\omega^2 + 4k_m D_s]^{1/2}}{2D_s}x\right\} - \left[\frac{FkP_0\omega^2}{D_s k^2 + (1 + K_P)\omega^2 k - k_m\omega^2}\right]\exp[(-k/\omega)x] + C_{eq}. \tag{6-64}$$

Equation (6-64) plots, in general, as two concave-down exponentials, one atop the other, which result in a concentration maximum (see Figure

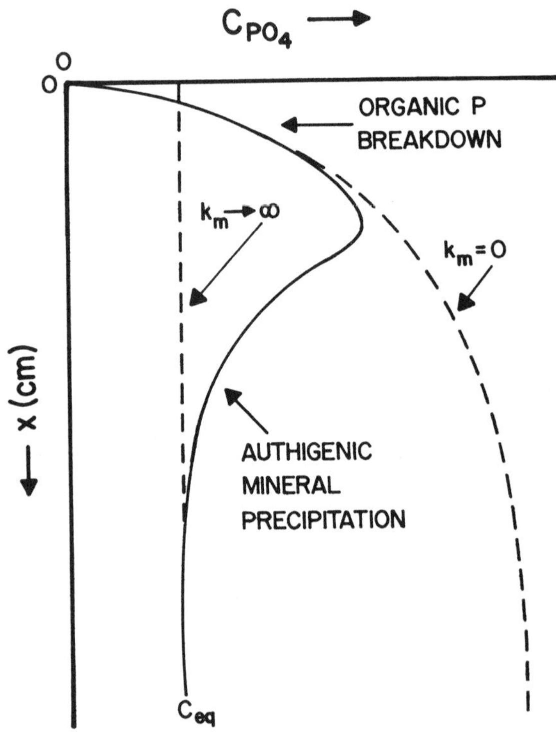

FIGURE 6-11. Schematic representation for general situation of steady state phosphate diagenesis according to equation (6-64). The special cases where $k_m \to \infty$ and $k_m = 0$ are shown by dashed lines.

6-11). So far, to the writer's knowledge, application of (6-64) to a specific sedimentary situation has not been done, because of a lack of independent knowledge of a sufficient number of the various parameters so that the remaining ones could be obtained by curve fitting. This is especially true of k_m because of a lack of precipitation rate data. However, Berner (1974) has used equation (6-64) in two special cases. They are (1) very rapid precipitation (high k_m) and (2) no phosphate precipitation ($k_m = 0$): In the first case as $k_m \rightarrow \infty$, equation (6-64) reduces to:

$$C = C_{eq}. \tag{6-65}$$

In other words, the complex kinetic expression is replaced by a simple equilibrium model. The resultant curve is shown in Figure 6-11. In the second case, if $k_m = 0$ (and letting $C = C_0$ at $x = 0$):

$$C = \left[\frac{FP_0\omega^2}{D_s k + (1 + K_P)\omega^2}\right]\{1 - \exp[(-k/\omega)x]\} + C_0. \tag{6-66}$$

This expression is identical to that for ammonia and applies when there is no precipitation. It is also shown in Figure 6-11.

Rapid precipitation of phosphate to form apatite is favored by the presence of fine-grained particles of calcium carbonate whose surfaces act as a nucleating agent for apatite crystallization (Stumm and Leckie, 1970; de Kanel and Morse, 1978). Otherwise, in the absence of appreciable $CaCO_3$ surface, the precipitation of apatite is strongly hindered by Mg^{++} in seawater (Martens and Harriss, 1970), and large degrees of supersaturation of interstitial water result. Vivianite precipitation is kinetically much easier but most *marine* pore waters do not attain sufficiently high concentrations of Fe^{++} and PO_4^{---} for supersaturation to occur. Thus, one would expect to find situation 1 above (for saturation with apatite) in fine-grained carbonate sediments and situation 2 (no precipitation) for other sediments. Vivianite precipitation is strongly suggested for the Sachem Harbor sediments discussed above (under ammonia), but these sediments are unusually enriched in highly reactive organic matter which enables all sulfate to be reduced, all dissolved sulfide to be removed, and consequently for Fe^{++} to build up to high concentrations along with phosphate.

Saturation of interstitial water of the fine-grained carbonate sediments of Bermuda and Florida Bay with respect to apatite has been demonstrated (Berner, 1974) and results are shown in Table 6-4. Concentrations are much lower than those found in non-carbonate sediments of similar

TABLE 6-4

Measured dissolved phosphate concentrations and values calculated (from pH, Ca^{++}, and F^- analyses) for equilibrium with marine apatite (phosphorite nodules) in fine-grained $CaCO_3$ sediments of Bermuda and Florida Bay. $\sum PO_4$ refers to total dissolved phosphate (i.e., $H_2PO_4^- + HPO_4^{--} + PO_4^{---}$ + ion pairs). (Data from Berner, 1974.)

pH	*μmoles per liter* $\sum PO_4$ *(calc.)*	$\sum PO_4$ *(meas.)*
	Bermuda	
7.45	2.0	1.2
7.40	2.3	1.8
7.38	2.5	1.7
7.42	2.3	1.7
7.53	2.7	3.4
7.48	3.2	4.2
7.21	7.9	9.1
7.19	8.3	6.7
6.85	26	32
6.73	38	48
	Florida Bay	
7.25	8.0	6.1
7.28	8.0	3.3
7.29	7.4	3.0
7.19	10	18
7.13	12	19
7.15	11	21
7.16	11	26
7.18	9	25

organic matter concentration, and they strongly suggest apatite precipitation as the dominant early diagenetic process controlling phosphate concentrations.

Situation 2, where there is no precipitation, is illustrated by a core from the FOAM site of Long Island Sound. Berner (1977) has fitted the following empirical expression (below the zone of bioturbation) to the data of Goldhaber et al. (1977):

$$C = 0.15\,(1 - \exp[-0.012x]) + 0.20, \qquad (6\text{-}67)$$

where C is in μmole cm^{-3}. Identification of this expression with equation (6-66) above, yields:

$$\frac{k}{\omega} = 0.012 \text{ cm}^{-1}, \tag{6-68}$$

$$\frac{FP_0\omega^2}{D_s k + (1 + K_P)\omega^2} = 0.15 \ \mu\text{mole cm}^{-3}. \tag{6-69}$$

Using the estimated value $\omega = 0.1 - 0.3$ cm yr^{-1} (Goldhaber et al., 1977; Rosenfeld, 1980) and the measured values for sediments from this site (Krom and Berner, 1980):

$$D_s = 100 \text{ cm}^2 \text{ yr}^{-1},$$
$$K_P = 2.0,$$

and the value of $F = 1.35$ gm cm^{-3}, we obtain:

$$k = 1.2 - 3.6 \times 10^{-3} \text{ yr}^{-1},$$
$$P_0 = 1.7 - 0.8 \ \mu\text{mole gm}^{-1},$$
$$= 0.0052 - 0.0024\text{‰}.$$

This value has not been checked via independent measurements of organic phosphorus. However, the carbon-to-phosphorus ratio of the decomposing organic matter can be deduced from the phosphorus analogue of equation (6-60):

$$\frac{G}{P} = \frac{-1}{\mathscr{L}}\left[\frac{\omega^2 + kD_{sS}}{\omega^2(1 + K_P) + kD_{sP}}\right]\frac{dC_S}{dC_P}, \tag{6-70}$$

where D_{sP} = diffusion coefficient of phosphate;
C_P = concentration of dissolved phosphate.

From the paper cited above (Berner, 1977) it is shown that $k_P = k_S$ $(=k_N) = k$ for one core (20D) at the FOAM site. Substituting the values: $k = 3.6 \times 10^{-3}$ yr^{-1}, $\omega = 0.3$ cm yr^{-1}, $D_{sS} = 100$ cm^2 yr^{-1}, $D_{sP} =$ 70 cm^2 yr^{-1}, $K_P = 2.0$, $\mathscr{L} = 0.5$, $dC_S/dC_P = -133$:

$$\frac{G}{P} = 230.$$

This ratio is considerably higher than the $C:P$ ratio of average marine plankton (Redfield, 1958) and suggests that either plankton of Long Island Sound are impoverished in phosphorus or selective loss of phosphorus relative to carbon occurs within or above the zone of bioturbation so that organic matter buried below it is low in phosphorus. Judging from the rapid rises in dissolved phosphate which occur within the zone of bioturbation in these sediments the author feels that the second explanation is more nearly correct. This is supported by preliminary analyses of Long Island Sound plankton (M. Krom, personal communication) which indicate a normal $C:P$ ratio of about 100.

Methane Formation

In many organic-rich sediments, both marine and non-marine, high concentrations of dissolved methane and even bubbles of gaseous methane are found. The methane forms microbiologically from the breakdown of organic matter via a sequence of fermentation reactions (e.g., see Claypool and Kaplan, 1974). Methane is common at all depths in anoxic freshwater sediments but, in marine sediments, it builds up to appreciable concentrations only at depths where sulfate has been completely removed by bacterial sulfate reduction (Barnes and Goldberg, 1976; Reeburgh and Heggie, 1974; Martens and Berner, 1977). The explanation usually offered for this observation is that in the presence of sulfate any methane produced locally or diffusing up from below is consumed by sulfate-reducing bacteria and associated fermentative microorganisms. In this way methane concentrations are kept at low levels within the zone of sulfate reduction, whereas they may build up at greater depths where sulfate is absent. In freshwater sediments sulfate concentrations at the time of burial are much lower because of much lower salinities. Consequently, complete sulfate reduction occurs quickly during diagenesis enabling a considerable build-up of methane near the sediment-water interface.

Evidence for direct methane consumption (oxidation) by sulfate-reducing bacteria in the laboratory is sparse (e.g., Sorokin, 1957), and, as a result, it was formerly believed that methane is not consumed anoxically. (By contrast, oxidation of methane by O_2 is well documented.) However, diagenetic modeling of pore water data in continental margin sediments provides direct evidence for consumption during sulfate reduction. For example, Barnes and Goldberg (1976) have shown that the concentration-vs-depth curves for dissolved methane found in sediments of the Santa Barbara Basin, California, can be explained only in terms of removal of methane within the zone of sulfate reduction. The methane concentration profile is strongly concave-up within the zone of sulfate reduction, which is the general situation found in other marine sediments. They believed that this shape could not result if there were methane production, without consumption, within the sulfate zone (concave-down), or production below the sulfate zone with upward diffusion through this zone (straight line).

The latter argument has also been advanced forcefully by Reeburgh and Heggie (1974). However, because of deposition, it is possible that diffusion through the zone of sulfate reduction, without reaction, could result in a concave-up concentration profile. The reason for this is that downward burial of sediment is opposed to the upward diffusion of methane or, in other words, the sediment-water interface (where methane concentration is fixed at a low level by aerobic oxidation) is not stationary relative to layers undergoing burial.

To test the idea that deposition of sediment could bring about the upward concavity for methane found at the FOAM site, Martens and Berner (1977) constructed a diagenetic model. It was assumed that only molecular diffusion and burial advection occur between the depth where methane reaches saturation (see Figure 6-12) and the sediment surface. Under these conditions the appropriate diagenetic equation, for steady state is simply:

$$D_s \frac{\partial^2 C}{\partial x^2} - \omega \frac{\partial C}{\partial x} = 0, \qquad (6\text{-}71)$$

where C = concentration of dissolved methane.

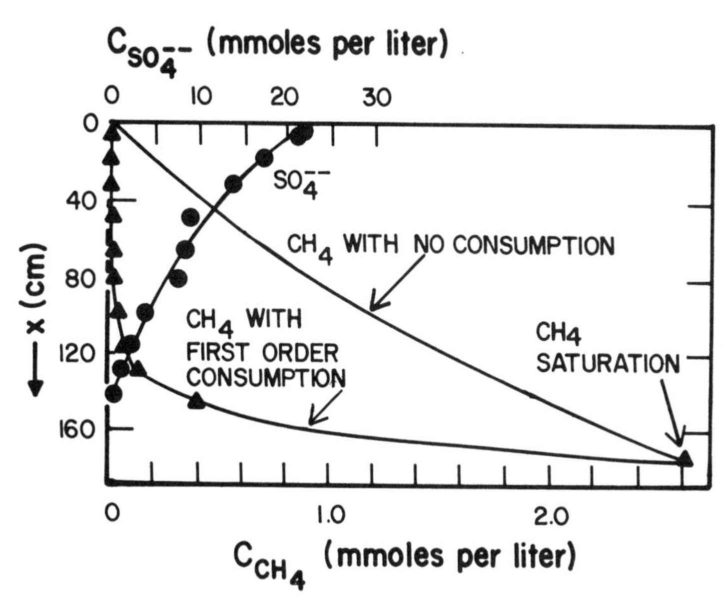

FIGURE 6-12. Measured data for dissolved methane and sulfate for sediments of the FOAM site of Long Island Sound. The methane data is well fitted by a theoretical curve calculated for first-order methane consumption. (After Martens and Berner, 1977.)

The boundary conditions used were:

$$x = 0; \qquad C = 0,$$

$$x = X; \qquad C = C_X.$$

Here C_X is the concentration of dissolved methane for saturation with respect to gaseous methane (e.g., it represents the depth where methane bubbles appear). The solution of (6-71) with these boundary conditions is:

$$C = \left[\frac{\exp(\omega/D_s)x - 1}{\exp(\omega/D_s)X - 1}\right] C_X. \tag{6-72}$$

For the approximate values of ω (0.3 cm yr^{-1}) and D_s (2×10^{-6} cm^2 sec^{-1}) appropriate for dissolved methane at the FOAM site, a curve is plotted in Figure 6-12. Note that the theoretical curve comes nowhere near fitting the measured concentrations and, in fact, is very close to being a straight line. This indicates that upward motion of the sediment-water interface due to deposition cannot be used to explain the strong upward concavity of the measured profile.

Martens and Berner then constructed diagenetic equations which assume consumption of methane within the zone of sulfate reduction. Both zero order and first order rate laws were tested. The measured data could not be fitted by *any* zero order model. The diagenetic equation with first order consumption is:

$$D_s \frac{\partial^2 C}{\partial x^2} - \omega \frac{\partial C}{\partial x} - k_M C = 0, \tag{6-73}$$

where k_M = first order removal rate constant for methane. Solution of (6-73) for the same boundary conditions given above yields:

$$C = \left[\frac{\exp(\alpha x) - \exp(\gamma x)}{\exp(\alpha X) - \exp(\gamma X)}\right] C_X, \tag{6-74}$$

where

$$\alpha = \frac{\omega + (\omega^2 + 4k_M D_s)^{1/2}}{2D_s};$$

$$\gamma = \frac{\omega - (\omega^2 + 4k_M D_s)^{1/2}}{2D_s}.$$

Equation (6-74) can be well fitted to the measured methane concentrations as shown in Figure 6-12, using the values given above for D_s and and ω, and the value:

$$k_M = 0.24 \text{ yr}^{-1}.$$

This value, when combined with the concentrations of methane shown in Figure 6-12, results in calculated methane consumption rates which are of the same order of magnitude as the rates of organic matter consumption calculated via the one-G model for bacterial sulfate reduction in the same sediment. Thus, this value of k_M is reasonable. Also, it is low enough to explain the lack of observation of methane consumption during sulfate reduction in the laboratory. At this rate of reaction a time scale of several months would be required to observe appreciable changes in methane concentration, and laboratory experiments are not normally conducted for such long periods.

These calculations show how a diagenetic bacterial reaction (anoxic methane consumption) which is too slow for study by means of laboratory experiments can be elucidated in natural sediments. Although an improved rate law, which takes into consideration the concentrations of both methane and sulfate, is needed, the preliminary calculations do illustrate the power of diagenetic modeling as a geochemical tool.

7

Pelagic (Deep-Sea) Sediments

Pelagic sediments (sometimes called deep-sea sediments) are here defined as those deposited far from the continents, usually in deep water (3,000–6,000 m), at rates slower than 10 cm per thousand years. These sediments are characterized by relatively low organic contents. Consequently, organic matter does not play as dominant a role during early diagenesis as it does for sediments of the continental margins, and non-biological processes such as mineral dissolution become quantitatively more important. Some outstanding examples of early diagenetic processes in pelagic sediments are: the dissolution of $CaCO_3$, the dissolution of opaline silica, the formation of ferromanganese nodules, and the formation of clays and zeolites from the reaction of volcanic ash or basalt with seawater. In those sediments that contain sufficient organic matter one also encounters altered pore water composition representing the early (suboxic) stages of organic matter decomposition, i.e., deoxygenation, nitrification, denitrification, and iron and manganese oxide reduction.

Pelagic sediments have been extensively sampled at considerable depth as a part of the Deep Sea Drilling Program. As a result, it is possible to follow diagenetic changes in them as they occur over hundreds of meters as opposed to the usual few meters or less sampled by ordinary coring. Also, through the use of pore water chemistry of these sediments, which has been determined at several locations, it is possible to construct various models to describe and explain diagenetic changes over moderately large depths and time scales (tens of millions of years).

In this chapter our goal will be to show how diagenetic modeling can be applied to the study of early diagenesis in fine-grained pelagic sediments. Topics covered include: calcium carbonate dissolution, opaline silica dissolution, suboxic organic diagenesis, redistribution of radium (as an example of radioisotope diagenesis), basalt-seawater reaction, and the effects of compaction, due to deep burial, upon diffusion.

Calcium Carbonate Dissolution

Calcium carbonate, as calcite, is a very common and abundant constituent of pelagic sediments. It is derived from the settling out of the skeletal remains of planktonic organisms, chiefly foraminifera (animals) and coccolithophorids (plants). (Aragonite in the form of pteropod remains is also a common constituent of calcareous material sedimented to the bottom, but because of its greater solubility, it disappears at relatively shallow depths and is, consequently, only a minor constituent of pelagic sediments.)

Although common, calcium carbonate is found only at the shallower oceanic depths, disappearing rapidly below about 4,000–5,000 meters. The reason for this disappearance is that at the greater depths calcite is removed by dissolution before it can become buried in the sediment. Although surface seawater is supersaturated with respect to calcite (and aragonite), at depths below about 500 meters in the Pacific and 2,000 meters in the Atlantic (Takahashi, 1975) the water becomes undersaturated and the degree of undersaturation increases with increasing depth (see Figure 7-1).

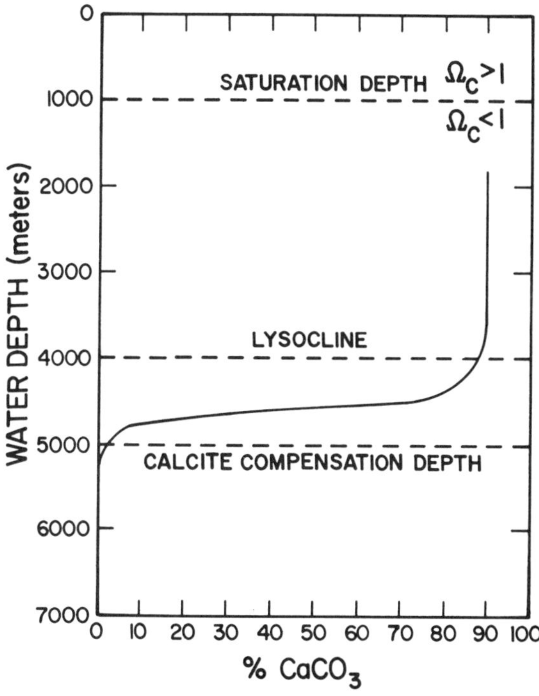

FIGURE 7-1. Generalized plot of percent calcium carbonate in bottom sediments vs water depth for the Pacific Ocean. Note the rapid change in calcium carbonate content between the lysocline and compensation depth.

Material settling to the bottom is exposed, while resting on the bottom, to more and more undersaturated water with increasing water depth, and, thus, faster and faster dissolution rates until a depth is reached where the rate of dissolution is equal to the rate of supply from above. Below this depth, called the carbonate compensation depth (see Figure 7-1), calcite disappears due to dissolution. Evidence for partial dissolution, in the form of the disappearance of more reactive shells or portions of shells, (due to higher surface energy, etc.) can be seen at depths considerably shallower than the compensation depth. This gives rise to the concept of the lysocline (Berger, 1970) or depth where evidence for considerable (selective) dissolution is first encountered. The lysocline is, on the average, about 1,000 meters shallower than the carbonate compensation depth, and most dissolution occurs within this 1,000-meter interval. Further general discussion of the nature and origin of the lysocline and compensation depth and the chemical and oceanographic factors bringing them about is beyond the scope of the present book. The interested reader is referred to the recent compendia of Andersen and Malahoff (1976), Berger (1976), and Morse and Berner (1979) for further information.

If most calcium carbonate dissolution occurs while the skeletal particles are resting on the bottom, and not while they are settling out, then the process falls within the domain of early diagenesis. This is generally believed to be the case, and as a result, has given rise to several diagenetic models for calcium carbonate dissolution (Schink and Guinasso, 1977; Takahashi and Broecker, 1977; Keir, 1979). Our purpose here will be to develop a very general model and then to show how the simpler models of these investigators are related to it. The ultimate goal of such modeling is to be able to predict, for given rates of sedimentation of $CaCO_3$ and clay, and a given relation between degree of undersaturation and water depth, at what water depth the $CaCO_3$ will be completely removed via dissolution during early diagenesis. In other word, the aim is to be able to predict the compensation depth. Because of the complexity of this problem and the considerable attention it has received the discussion here will be rather lengthy.

A general model for $CaCO_3$ dissolution during early diagenesis must take into consideration the following factors: molecular diffusion, bioturbation, depositional burial, compaction and flow as a result of water loss, compaction as a result of the loss of $CaCO_3$ by dissolution, equilibrium adsorption of Ca^{++}, the kinetics of $CaCO_3$ dissolution, the rate of input to the sediment surface of both $CaCO_3$ and insoluble material ("clay"), and the saturation state of the bottom water immediately overlying the sediment. Because of reasons given in Chapter 3 the effects of a possible diffusive boundary layer overlying the sediment will be ignored. Of the factors listed above, the one most unique to this problem is compaction brought about by the dissolution, and because this is basically a new diagenetic process not previously discussed in this book, it merits special attention.

Compaction of sediment is not necessarily brought about by pushing of clay platelets or other particles close together. Compaction can also occur if particles within the sediment pass into solution. A simple way to visualize this process is as a stack of dirty snowballs immersed in alcohol (Figure 7-2). Assume that snowballs at the bottom of the stack melt and the water mixes with the alcohol. Melting of the bottom snowballs causes the stack to collapse, resulting in squirting upward of the water-alcohol mixture. To maintain steady state, new snowballs are added at the top to replace those lost by melting, and meltwater is ejected to the overlying alcohol. Also, dirt left behind by melting accumulates at the bottom of the stack. This model is analogous to $CaCO_3$ dissolution in a sediment. The snowballs represent $CaCO_3$, the alcohol-water the interstitial solution containing dissolved $CaCO_3$, and the dirt the residual insoluble clay. Complete dissolution is what is represented in Figure 7-2. If the interstitial solution instead came to equilibrium with $CaCO_3$ (or ice) before complete dissolution, there would be some $CaCO_3$ (snowballs) buried with the clay (dirt).

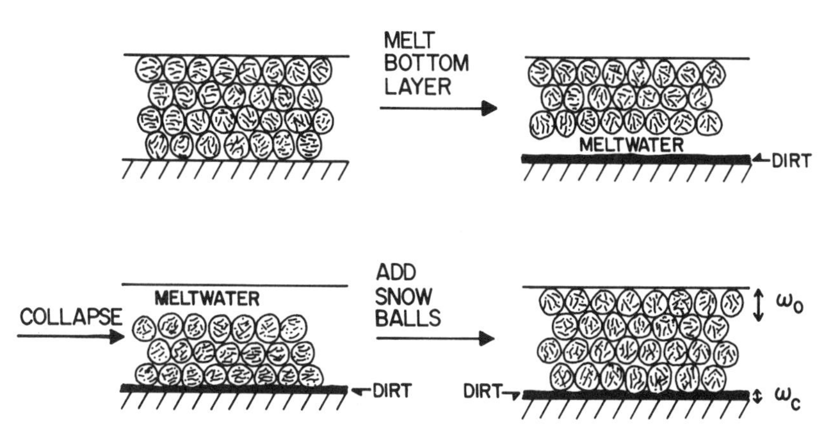

FIGURE 7-2. Schematic representation of the "dirty snowball" analogue for compaction due to $CaCO_3$ dissolution in pelagic sediments. Upon completing the cycle of melting, collapse, squirting upward of meltwater, and addition of new snowballs, the rate of burial can be seen to be higher at the top (ω_0) where snowballs are buried, than at the bottom (ω_c), where only dirt is buried (e.g., see Farrow, 1953.)

The effects of dissolution-induced compaction can be expressed mathematically. From equations (3-1) and (3-2) of Chapter 3, the mass balance expressions for total solids ($CaCO_3$ plus clay) and interstitial water are respectively:

$$\frac{\partial[(1-\phi)\bar{\rho}_s]}{\partial t} = \frac{\partial\left\{D_B \frac{\partial[(1-\phi)\bar{\rho}_s]}{\partial x}\right\}}{\partial x} - \frac{\partial[(1-\phi)\bar{\rho}_s\omega]}{\partial x} - \hat{R}_d, \quad (7\text{-}1)$$

$$\frac{\partial(\phi\rho_w^*)}{\partial t} = \frac{\partial\left[D_B \frac{\partial(\phi\rho_w^*)}{\partial x}\right]}{\partial x} - \frac{\partial(\phi\rho_w^* v)}{\partial x}, \quad (7\text{-}2)$$

where $\hat{R}_d$ = rate of dissolution of $CaCO_3$ in mass per unit volume of total sediment (solids plus water);
$\bar{\rho}_s$ = average density of total solids;
ρ_w^* = mass of water per unit volume of pore solution (not density);
D_B = biodiffusion coefficient

Here the term $\hat{R}_d$ expresses the effect of dissolution on compaction. These equations can be combined with analogous expressions for solid $CaCO_3$ and dissolved $CaCO_3$ (as represented by dissolved CO_3^{--}). In the most general form, equations (3-73) and (3-74) are:

For solid $CaCO_3$ *(calcite):*

$$\frac{\partial[(1-\phi)\bar{\rho}_s N]}{\partial t} = \frac{\partial\left\{D_B \frac{\partial[(1-\phi)\bar{\rho}_s N]}{\partial x}\right\}}{\partial x} - \frac{\partial[(1-\phi)\bar{\rho}_s N\omega]}{\partial x} - \hat{R}_d. \quad (7\text{-}3)$$

For dissolved $CaCO_3$ (CO_3^{--})*:*

$$\frac{\partial(\phi C)}{\partial t} = \frac{\partial\left[D_B \frac{\partial(\phi C)}{\partial x} + \phi(D_s + D_I)\frac{\partial C}{\partial x}\right]}{\partial x} - \frac{\partial(\phi v C)}{\partial x} + \hat{R}_d + \sum\hat{R}_i, \quad (7\text{-}4)$$

where N = mass fraction of total solids present as calcium carbonate;
C = concentration of CO_{3aq}^{--} in mass per unit volume of pore water;
$\sum\hat{R}_i$ = all reactions affecting dissolved Ca^{++} other than dissolution (e.g., adsorption);
D_I = irrigation coefficient.

These equations can be considerably simplified using certain reasonable assumptions. They are:

1. The average density of solids, $\bar{\rho}_s$, does not change (with depth or time) as a result of deposition or dissolution. This is equivalent to stating that the densities of clay and $CaCO_3$ are essentially the same.
2. The mass of water per unit volume of interstitial solution ρ_w^* does not change with depth or time. This is equivalent to saying that the addition of $CaCO_3$ to solution does not appreciably dilute the water in solution.
3. Porosity does not change as a result of dissolution. This is a more restricted assumption but is not unreasonable since $CaCO_3$-rich and $CaCO_3$-poor sediments in the deep sea have approximately the same porosity (see Keir, 1979).
4. Compaction due to collapse of clay floccules (normal compaction) is negligible and there is no externally impressed flow.
5. Reactions affecting dissolved Ca^{++} and CO_3^{--}, other than dissolution, are unimportant. In other words, adsorption of Ca^{++} is neglected.
6. Molecular diffusion is much more rapid than pore water bioturbation (good assumption for pelagic sediments).
7. The particle biodiffusion coefficient is constant over the zone of dissolution.
8. Steady state diagenesis is present.

As a consequence of (1), (2), (3), (4), and (8):

$$\left.\begin{aligned} \frac{\partial \bar{\rho}_s}{\partial x} = \frac{\partial \rho_w^*}{\partial x} = \frac{\partial \phi}{\partial x} = 0, \\ \frac{\partial \bar{\rho}_s}{\partial t} = \frac{\partial \rho_w^*}{\partial t} = \frac{\partial \phi}{\partial t} = \frac{\partial N}{\partial t} = \frac{\partial C}{\partial t} = 0. \end{aligned}\right\} \tag{7-5}$$

Using these assumptions, equations (7-1)–(7-4) reduce to:

$$\frac{\partial \omega}{\partial x} = -R_d', \tag{7-6}$$

$$\frac{\partial v}{\partial x} = 0, \tag{7-7}$$

$$D_B \frac{\partial^2 N}{\partial x^2} - \omega \frac{\partial N}{\partial x} - N \frac{\partial \omega}{\partial x} - R_d' = 0, \tag{7-8}$$

$$D_s \frac{\partial^2 C}{\partial x^2} - v \frac{\partial C}{\partial x} - C \frac{\partial v}{\partial x} + FR_d' = 0, \tag{7-9}$$

where

$$R'_d = \hat{R}_d/(1 - \phi)\bar{\rho}_s;$$
$$F = (1 - \phi)\bar{\rho}_s/\phi.$$

From (7-6):

$$\omega = \omega_f + \int_x^{x_f} R'_d\,dx, \qquad (7\text{-}10)$$

where ω_f refers to the final rate of burial (after dissolution) as normally measured, and x_f is the depth where saturation (or complete dissolution) is attained. Because of a lack of porosity change $v = \omega_f$.

Substituting (7-6), (7-7), and (7-10) in (7-8) and (7-9) we obtain finally:

$$D_B\frac{\partial^2 N}{\partial x^2} - \left[\omega_f + \int_x^{x_f} R'_d\,dx\right]\frac{\partial N}{\partial x} - (1 - N)R'_d = 0, \qquad (7\text{-}11)$$

$$D_s\frac{\partial^2 C}{\partial x^2} - \omega_f\frac{\partial C}{\partial x} + FR'_d = 0. \qquad (7\text{-}12)$$

Equations (7-11) and (7-12) can be further simplified if the rate expressions for dissolution are known. Experimental work of Morse (1978) and Keir (1979) has demonstrated that the dissolution of calcite as forams, coccoliths, reagent calcite, and natural sediments in seawater can be represented by the expression:

$$\bar{R} = k(1 - \Omega_c)^n, \qquad (7\text{-}13)$$

where $\bar{R}$ = rate of dissolution in mass per unit surface area of calcite per unit time;

k = rate constant or maximum rate of dissolution (when $\Omega_c = 0$);

n = an empirical constant ($= 4 \pm 1$);

Ω_c = the degree of saturation with calcite (see Chapter 4).

In seawater the concentration of Ca^{++} is much greater than that of CO_3^{--} and it remains roughly constant upon equilibration with calcite. In this case:

$$(1 - \Omega_c) \approx \frac{(C_{eq} - C)}{C_{eq}}. \qquad (7\text{-}14)$$

The measured parameter $\bar{R}$ is related to R'_d as:

$$R'_d = NA_v\bar{R}, \qquad (7\text{-}15)$$

where A_v = specific surface area of calcite (area per unit mass of calcite). Upon substitution of (7-13), (7-14), and (7-15) in (7-11) and (7-12), we obtain:

$$D_B \frac{\partial^2 N}{\partial x^2} - \left[\omega_f + \int_x^{x_f} \frac{N A_v k}{C_{eq}^n}(C_{eq} - C)^n \, dx\right] \frac{\partial N}{\partial x} - \frac{(1 - N) N A_v k}{C_{eq}^n}(C_{eq} - C)^n = 0, \quad (7\text{-}16)$$

$$D_s \frac{\partial^2 C}{\partial x^2} - \omega_f \frac{\partial C}{\partial x} + \frac{F N A_v k}{C_{eq}^n}(C_{eq} - C)^n = 0. \quad (7\text{-}17)$$

Equations (7-16) and (7-17) are quite general in scope but are not easily solved by analytical methods. They will not be solved here. Instead, they will be used to show how the simpler diagenetic models of Schink and Guinasso (1977) and Keir (1979) are related to the model presented here. The model of Schink and Guinasso can be derived directly from (7-16) and (7-17) by further assuming:

1. Change in ω with depth, due to dissolutive compaction, is negligible compared to changes in the concentration of $CaCO_3$. In other words,

$$\omega \frac{\partial N}{\partial x} \gg N \frac{\partial \omega}{\partial x} \quad \text{so that} \quad \frac{\partial (N\omega)}{\partial x} \approx \omega \frac{\partial N}{\partial x}.$$

2. Dissolution proceeds via linear kinetics, i.e., $n = 1$.
3. The surface area of dissolving calcite is directly proportional to the mass present in the sediment. In other words, specific surface area of calcite, A_v, is constant during dissolution.

With these assumptions, and upon substituting $B = \bar{\rho}_s(1 - \phi)N$, equations (7-16) and (7-17) simplify to:

$$D_B \frac{\partial^2 B}{\partial x^2} - \omega_f \frac{\partial B}{\partial x} - \frac{k_B B}{C_{eq}}(C_{eq} - C) = 0, \quad (7\text{-}18)$$

$$D_s \frac{\partial^2 C}{\partial x^2} - \omega_f \frac{\partial C}{\partial x} + \frac{k_B B}{\phi C_{eq}}(C_{eq} - C) = 0, \quad (7\text{-}19)$$

where $k_B = A_v k$ = constant. Here B represents the mass of calcium carbonate per unit volume of total sediment (solids plus water).

To solve (7-18) Schink and Guinasso used the following boundary conditions for B:

$$x = 0; \qquad F_B = -D_B \frac{\partial B}{\partial x} + \omega_f B,$$

$$x = x_f; \qquad \partial B/\partial x = 0,$$

where F_B = flux of $CaCO_3$ to the sediment-water interface by sedimentation. For C a more complicated upper boundary condition was used to allow for the (assumed) presence of a one millimeter thick diffusive boundary sublayer in the overlying water (see Chapter 3). Accordingly, the diffusive flux to the sediment-water interface was equated to that away from the interface within the diffusive sublayer:

$$x = 0; \qquad D\left(\frac{\partial C_w}{\partial x}\right)_0 = \phi D_s\left(\frac{\partial C}{\partial x}\right)_0,$$

where D = molecular diffusion coefficient in the overlying seawater;
C_w = concentration of dissolved CO_3^{--} in the overlying seawater.

The lower boundary condition for C was, as for B:

$$x = x_f; \qquad \partial C/\partial x = 0.$$

Schink and Guinasso, using these boundary conditions, solved (7-18) and (7-19) by a numerical method. For this purpose they used the values: $F_{clay} = 1.5 \times 10^{-4}$ gm cm^{-2} yr^{-1} (depositional flux of insoluble material), $\phi = 0.80$, $\bar{\rho}_s = 2.6$ gm cm^{-3}, $D_s = 240$ cm^2 yr^{-1}, $D_B = 0.3$ cm^2 yr^{-1}, and $C_0 = 0.1$ μmole cm^{-3} (concentration of dissolved CO_3^{--} in the overlying seawater). Influx rate of $CaCO_3(F_B)$ or dissolution rate constant (k_B) were allowed to vary, and C_{eq} was calculated from known values of calcite solubility in seawater as a function of water depth (pressure). From the results, curves were calculated of weight percent $CaCO_3$, in sediment buried below the zone of dissolution, vs water depth. Results are shown in Figure 7-3. As can be seen, the model predicts a sharp drop in percent $CaCO_3$ over a narrow depth range even though linear kinetics are assumed. This is the general situation actually found in the oceans (see Figure 7-1), so that as a first approximation, the model appears to be valid. Strong sensitivity to the rate of input, via deposition, of $CaCO_3$ is also shown in Figure 7-3a. Variation of other critical parameters was found to produce contrasting results. Sensitivity to changes in k_B (dissolution rate constant) of the percent $CaCO_3$-vs-depth curves (Figure 7-3b) is less

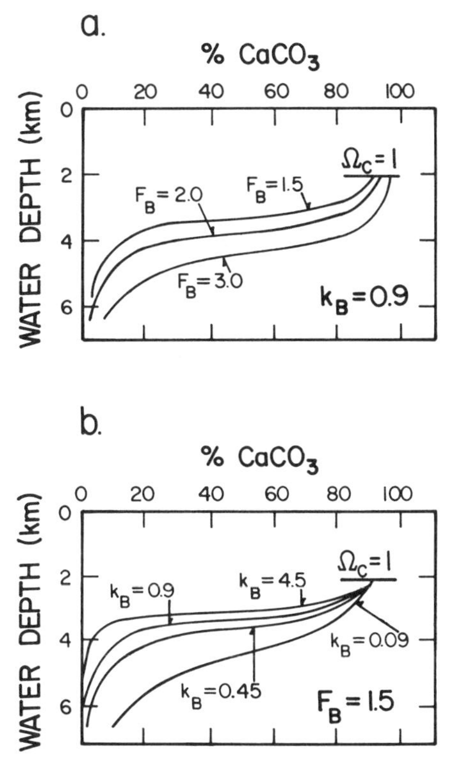

FIGURE 7-3. Theoretical plots of weight % $CaCO_3$ vs water depth for pelagic sediments. The parameter F_B represents the flux (in gm cm^{-2} kyr^{-1}) of sedimenting $CaCO_3$ to the sediment-water interface. The parameter k_B is the dissolution rate constant (in yr^{-1}). (After Schink and Guinasso, 1977.) Values of other parameters, fixed in both diagrams, are: $D_B = 0.3$ cm^2yr^{-1}, $D_s = 240$ cm^2yr^{-1}, $\phi = 0.8$, $\bar{\rho}_s = 2.6$ gm cm^{-3}, $F_{clay} = 1.5 \times 10^{-4}$ gm $cm^{-2}yr^{-1}$, $C_0 = 0.1$ μmol cm^{-3} (see text).

than sensitivity to changes in F_B. The effects of large changes in bioturbation rate (D_B) are minor, whereas changes in the thickness of the diffusive sublayer, brought about by small changes in bottom water flow velocity, result in significant shifts in the curves.

In the model of Keir (1979), compaction due to dissolution is not ignored, as was done by Schink and Guinasso (1977). However, Keir greatly simplifies equation (7-16) for solids by assuming that particle bioturbation is very fast relative to deposition and dissolution. In other words, Keir assumes a box model (see Chapter 3) for the solids over the top few centimeters of sediment where bioturbation is active. Within the box, concentration of solid $CaCO_3$ is assumed to be constant with depth (due to bioturbation) over the zone of dissolution (top 0.3 cm), whereas concentration of dissolved $CaCO_3$ continually changes with depth as predicted by equation (7-17). Besides the use of a box model for solids, Keir's other basic assumptions are:

1. Burial flux of total solids through the bottom of the box is less than depositional flux to its top. The difference is equal to the upward diffusive flux of dissolved $CaCO_3$ out of the sediment.
2. The advective term in equation (7-17) is small compared to the terms for diffusion and dissolution, and thus can be ignored.
3. The surface area of a dissolving calcite particle remains constant during dissolution. This means that for a constant initial size of calcite particles, the rate of dissolution is directly proportional to the *number* of particles present, and not to their total mass (as assumed by Schink and Guinasso). This is a reasonable approximation for thin spherical shells (forams) and thin discs (coccoliths). By this model the rate of dissolution

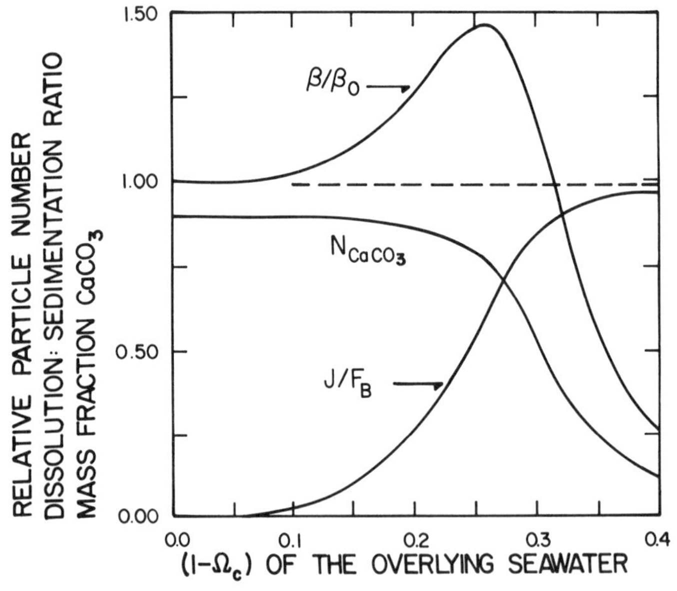

FIGURE 7-4. Plots of some critical parameters according to the steady-state particle-box model of Keir (1979). The ratio β/β_0 (relative particle number) represents the number of particles of calcite in the box divided by the number of particles of calcite that would have been present had there been no dissolution. The ratio J/F_B represents the ratio of dissolution flux J (migration of dissolved calcium carbonate upward out of the sediment via diffusion) to sedimentation flux of calcium carbonate F_B. Mass fraction of calcium carbonate (relative to total sediment solids) is denoted by the symbol N_{CaCO_3}. (After Keir, 1979.)

actually increases as calcite is consumed, due to initial increase in the number of particles in the box, before decreasing in later stages (see Figure 7-4).

4. Porosity ϕ and average density of solids $\bar{\rho}_s$ do not change as a result of dissolution.
5. Steady state diagenesis is present.

With these assumptions, Keir solved the appropriate equations for his particle-box model and, by a suitable choice of dissolution rate constants, could obtain good fits to actual measured percent $CaCO_3$-vs-depth data for various regions of the ocean bottom. An example is shown in Figure 7.5. Unfortunately, the curve-fit rate constants were found to be considerably lower than those determined in the laboratory by dissolving deep-sea calcareous sediments in undersaturated seawater. The discrepancy was explained by Keir as being due to dissolution-inhibiting factors present in natural sediments which are not taken into account in laboratory experiments.

One factor which has been postulated to explain inhibited dissolution in sediments (Schink and Guinasso, 1978b; see also Keir, 1979), is the

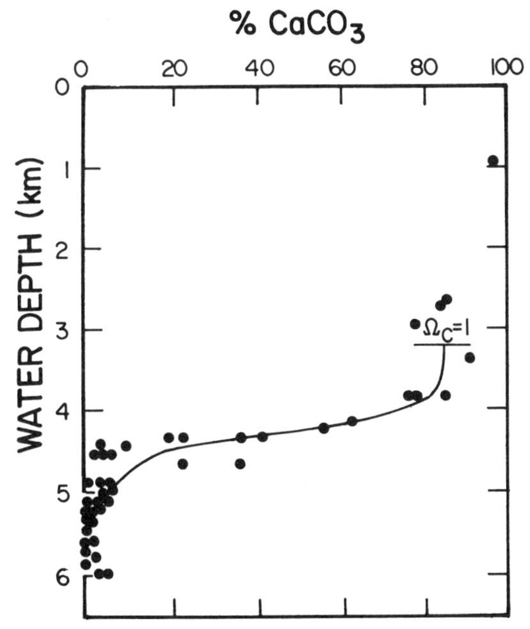

FIGURE 7-5. Plot of weight percent $CaCO_3$ vs water depth for sediments of the southwestern Atlantic Ocean. The curve is calculated via the particle-box model. (After Keir, 1979.)

presence of aragonite in the sediments. Aragonite in the form of planktonic snails, called pteropods, is known to sediment to the seafloor, and there rapidly to undergo dissolution. (Aragonite dissolution is faster than calcite dissolution, under the same conditions, because aragonite has a higher solubility). If the aragonite is mixed, by bioturbation, into the sediment, its continued dissolution could add dissolved $CaCO_3$ to the interstitial water of the sediment, and thus slow the rate of calcite dissolution by raising the value of C. Inclusion of aragonite dissolution in diagenetic equations by Schink and Guinasso (1978b) has led them to conclude that this process can, indeed, explain much of the rate inhibition found in deep-sea sediments as well as the separation of the lysocline from the much shallower depth where calcite becomes undersaturated in the oceans.

Opaline Silica Dissolution

Planktonic diatoms (plants) and radiolaria (animals) living in the surface waters of the oceans secrete skeletons consisting of opaline ("amorphous") silica. Because of the excess energy provided by sunlight, they are able to do this even though seawater is everywhere decidedly undersaturated with respect to opaline silica. Once they die, their siliceous remains settle to the bottom and undergo dissolution. Most of the dissolution occurs during settling (Broecker, 1971; Hurd, 1973) but some also occurs on the bottom. Evidence for dissolution on the bottom is provided by concentrations of dissolved silica in sediment pore water which are almost always higher than those in the overlying seawater. In some areas of the deep ocean, especially the Antarctic, dissolution is incomplete, and because of a large input of biogenic material, the opaline silica accumulates to form a sediment type referred to as siliceous ooze. This ooze owes its existence to the fact that opaline silica dissolution occurs via a surface-reaction rate-controlling mechanism.

The rate of dissolution of freshly killed siliceous plankton has been studied in the laboratory by Lawson et al. (1978). They discovered that the rate was much slower than that previously found in other laboratory studies for siliceous bottom sediments. The reason for this is that the bottom sediment, before use, had to be treated with strong acid to remove other phases (e.g., $CaCO_3$, manganese oxides, etc.). This acid treatment apparently affected the surface of the biogenic silica and greatly increased its reactivity. The rates measured by Lawson et. al. were shown by Berner (1978b) to be over 6 *orders of magnitude* slower than that predicted for rate-control by molecular diffusion. Thus, opaline silica dissolution is a surface-reaction controlled process. Increased reactivity after acid treatment suggests that the slow dissolution rate is due partly to surface-adsorbed inhibitors, especially magnesium (e.g. Hurd, 1973; Wollast, 1974), which are partly removed by the acid treatment. If opaline silica were to dissolve via molecular diffusion or other transport-controlled processes, it would completely dissolve away before it could accumulate on the bottom at the slow rates of deposition found for pelagic sediments. Thus, the very slow, inhibited rate of dissolution is what enables biogenic siliceous ooze to form in undersaturated bottom water which is otherwise capable of completely destroying it.

The data of Lawson et al. (1978) are insufficient to deduce the functional dependence of dissolution rate upon the degree of undersaturation. However, the earlier work of Hurd (1972), on acid-treated material, indicates a first-order dependence. If the functionality (as opposed to the absolute rate) found for acid-cleaned material can be applied to natural biogenic silica, then the rate law for a sediment should be of the form:

$$R = k_m \bar{A}(C_{eq} - C), \tag{7-20}$$

where C = concentration of dissolved silica;
C_{eq} = concentration of dissolved silica at saturation with biogenic opaline silica;
k_m = rate constant;
$\bar{A}$ = surface area of opaline silica per unit volume of pore water.

Schink et al. (1975) in their model of silica diagenesis, discussed below, have adopted this rate law but in the slightly modified form:

$$R = \frac{k'_m B}{C_{eq}}(C_{eq} - C), \tag{7-21}$$

where B = concentration of reactive biogenic silica (of constant specific surface area);
k'_m = rate constant.

Several models for the early diagenesis of silica have been proposed (e.g., Hurd, 1973; Wollast, 1974). The most complete model is that of Schink et al. (1975). Schink et al. treated dissolution in terms of "reactive" or soluble silica to differentiate it from other less reactive forms such as silicate minerals. Their reactive silica is, however, better viewed as that fraction of the biogenic opaline silica which undergoes dissolution during burial. This is in keeping with the observation that biogenic silica deposited in sediments exhibits a large range in reactivity, which is directly manifested by preferential dissolution of different species (e.g., Johnson, 1975). As a first approximation for the purpose of mathematical modeling, one can divide biogenic silica into two categories, reactive and non-reactive, and treat only the reactive fraction as undergoing dissolution. By this approach the reactive fraction may completely dissolve without depleting all of the siliceous material or attaining saturation with opaline silica. An approach of this type is necessary because of the finding that pore waters at depth in sediments high in diatom and radiolarian debris do not reach saturation with opaline silica as measured in the laboratory (Hurd, 1973). Instead, asymptotic concentrations are always lower than that predicted for saturation, and are highly variable from one sediment locality to another. The variability cannot be explained in terms of the formation of a surface phase (e.g., sepiolite) of lowered solubility on the biogenic silica because such a phase should give about the same asymptotic concentration (solubility) for all sediments.

The basic assumptions of the model of Schink et al. (1975) are:

1. Reactive silica dissolves at a rate proportional to the first power of the departure from saturation according to equation (7-21).
2. Specific surface area of silica does not change with depth so that the rate of dissolution is proportional to the concentration of reactive solid silica (equation 7-21).
3. Advection due to burial is small compared to reaction and diffusion and, thereby, can be neglected for pelagic sediments.
4. Compaction, water flow, etc. are absent.
5. Particle bioturbation occurs in a surface zone and can be described by a constant biodiffusion coefficient within this zone.
6. Pore water bioturbation is much slower than molecular diffusion.
7. Silica adsorption, even though it occurs, may be ignored because of other assumptions (negligible advection and steady state).
8. Diagenesis is at steady state.

With these assumptions, the appropriate diagenetic equations used by Schink et al. (1975) are:

For reactive opaline silica:

$$D_B \frac{\partial^2 B}{\partial x^2} - \frac{k'_m(C_{eq} - C)}{C_{eq}} B = 0. \tag{7-22}$$

For dissolved silica:

$$D_s \frac{\partial^2 C}{\partial x^2} + \frac{k'_m B}{\phi C_{eq}} (C_{eq} - C) = 0, \tag{7-23}$$

where B = concentration of *reactive* opaline silica in mass per unit volume of total sediment;

C = concentration of dissolved silica in mass per unit volume of pore water.

The upper boundary conditions adopted by Schink et al. are that the concentration of dissolved silica at the sediment-water interface is the same as that of the overlying seawater, i.e., $x = 0$, $C = C_0$, and that the depositional flux of silica to the bottom is equal to the bioturbational flux downward from the sediment-water interface, i.e., $x = 0$, $F_B = -D_B(\partial B/\partial x)$, where F_B is the depositional flux. (The latter condition is true only because we have ignored depositional burial.) The lower boundary conditions are that asymptotic values are approached; in other words, for large x:

$$\partial B/\partial x = 0,$$
$$\partial C/\partial x = 0,$$
$$C = C_\infty.$$

Actually, by their model, they should have also stated that at great depth $B = 0$.

Solution of (7-22) and (7-23) for these boundary conditions was done numerically and the results plotted graphically. For this purpose the following values were used: $D_s = 4 \times 10^{-6}$ cm^2 sec^{-1}, $C_{eq} = 1.0$ μmole cm^{-3}, $\phi = 0.8$. Values of the other parameters were constrained within certain ranges through the use of D_B values determined on other sediments, laboratory measurements of k_m' for material untreated by acids (Johnson, 1975), and values found for the asymptotic concentration C_∞ and the depth necessary to reach C_∞ in a wide variety of pelagic sediments. From this data they constructed a series of diagrams relating C_∞, D_B, k_m, and F_B. An example is shown in Figure 7-6.

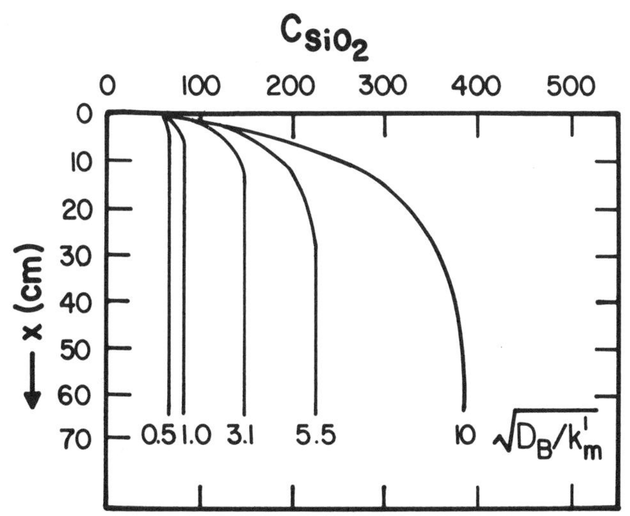

FIGURE 7-6. Plot of concentration of dissolved silica (in μmoles per liter) vs sediment depth as predicted by the diagenetic model of Schink et al. (1975). Note that (at constant k_m') increase in bioturbation rate D_B causes an increase in silica concentration at depth.

An interesting, and unexpected, result of the model calculations of Schink et al. is that increased bioturbation or decreased dissolution rate leads to an increase in the asymptotic concentration C_∞. This is shown in Figure 7-6. One might intuitively expect the opposite. The reason for this result is that increased bioturbation carries reactive material downward, where it can inject silica into solution at depths which, in the absence of bioturbation, would be below the depths of complete dissolution. Likewise, slower dissolution enables burial of reactive solid silica and injection at depth. Rapid injection near the surface leads to rapid loss of dissolved silica via molecular diffusion to the overlying water, and consequently to rapid disappearance of reactive solid silica before appreciable concentrations can build up in interstitial solution. This shows that diagenetic modeling can be used to point out new, unexpected relations as well as to calculate rates of diagenetic processes.

Although the model of Schink et al. leads to some interesting results, it poses a new problem namely, the nature of the material denoted by the symbol B. Is B a large fraction of the total biogenic silica or a small fraction? How does the reactivity of $B(k'_m)$ change from one area to another and especially from one organism (or group of organisms) to another? The latter question has been addressed qualitatively by Johnson (1975). What is needed now is more data on rates of dissolution of the actual materials involved in early diagenesis on the deep-sea floor.

Suboxic Organic Matter Diagenesis

Although pelagic sediments are low in organic matter, they do show some evidence of organic matter decomposition in the form of altered concentrations of certain ions in interstitial solution. For example, organic decomposition often results in the exhaustion of dissolved oxygen from the pore water. However, because bacterial reactions are very slow, we rarely encounter evidence for sulfate reduction within the top few meters of sediment. Instead, there are relatively thick zones representing earlier processes. These are: oxygen depletion accompanied by nitrate build-up (the same process that occurs in the overlying water), nitrate reduction, manganese oxide reduction, and ferric oxide reduction. The term "suboxic" diagenesis (Froelich et al., 1979) has been applied to these processes. They are normally missed in continental margin sediments because, due to rapid bacterial activity, the zones representing each process are extremely thin and masked in the top few meters of sediment by sulfate reduction.

Based on the free energy yield accompanying organic matter oxidation, one may expect a definite succession of suboxic processes during early diagenesis. This succession is summarized in Chapter 4 (Table 4-4) and from it one can predict a pattern of successive concentration changes vs

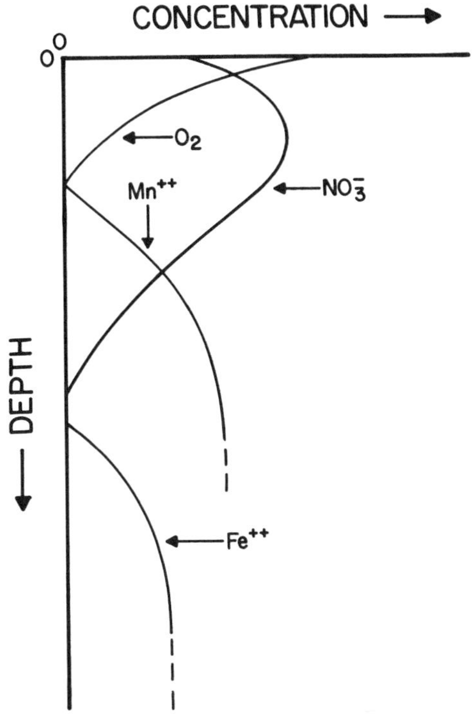

FIGURE 7-7. Schematic representation of depth trends in concentrations of dissolved species found in pelagic sediments. The succession of processes is: O_2 reduction, nitrification (nitrate formation), denitrification (nitrate reduction), MnO_2 reduction, and FeOOH reduction. (Adapted from Froelich et al., 1979.)

depth. Actual observations by Froelich et al. (1979), shown in a generalized form in Figure 7-7, agree well with the predicted pattern.

The data given by Froelich et al. for nitrate can be modeled diagenetically. An example is shown in Figure 7-8. If one assumes that nitrate reduction takes place via first order kinetics, by analogy with continental margin sediments as discussed in the previous chapter, it is possible to fit a diagenetic equation to the nitrate data and to obtain a value of the rate constant for nitrate reduction. If we also assume:

1. No bioturbation (nitrate reduction occurs at depth),
2. No reactions involving nitrate other than denitrification (reasonable in the absence of O_2),
3. No adsorption of NO_3^- (which is reasonable for this anion),
4. No compaction, water flow, etc.,
5. Steady state diagenesis,

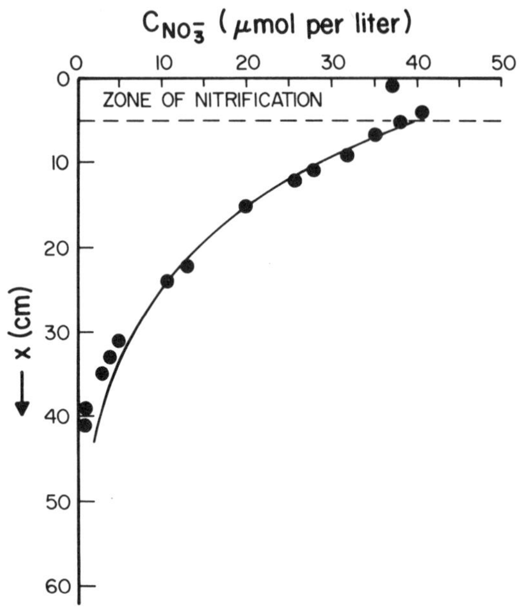

FIGURE 7-8. Plot of measured dissolved nitrate concentration vs depth for pelagic sediment 23 GCl from the eastern equatorial Atlantic Ocean. A theoretical curve, predicted by equation (7-26), is fitted to the data below 5 cm (zone of denitrification). (Data from Froelich et al., 1979.)

we obtain the simple diagenetic equation:

$$D_s \frac{\partial^2 C}{\partial x^2} - \omega \frac{\partial C}{\partial x} - k_{NO_3^-} C = 0, \tag{7-24}$$

where $k_{NO_3^-}$ = first order denitrification rate constant;

C = concentration of dissolved nitrate.

Boundary conditions for solution of (7-24) are a bit unusual and require discussion. At the maximum in C we have a balance between NO_3^- production by aerobic respiration (nitrification) and NO_3^- reduction (denitrification). We will assume that this maximum is maintained and that above it there is no denitrification and below it there is no nitrification. This is necessitated by the assumptions given above. Thus, our upper boundary condition is:

$$x = x_m,$$
$$C = C_m,$$

where the subscript m refers to the maximum. At depth nitrate is entirely depleted, but the reacting organic matter is not. In fact, the assumed independence of rate of reduction from organic matter concentration means that organic matter is present to large excess. This is reasonable for a minor species like nitrate (Vanderborght et al., 1977) as pointed out in Chapter 4, but not for sulfate where a G-model of some sort is more appropriate. At any rate, the lower boundary condition here is:

$$x \to \infty,$$
$$C \to 0.$$

Solution of (7-24) for these boundary conditions yields:

$$C = C_m \exp\left\{\left[\frac{\omega - (\omega^2 + 4k_{NO_3^-}D_s)^{1/2}}{2D_s}\right](x - x_m)\right\}. \qquad (7\text{-}25)$$

At the slow rates of deposition found in the deep sea:

$$4k_{NO_3^-}D_s \gg \omega^2.$$

Thus, equation (7-25) can be simplified to:

$$C = C_m \exp\left[-\left(\frac{k_{NO_3^-}}{D_s}\right)^{1/2}(x - x_m)\right]. \qquad (7\text{-}26)$$

This equation has been fitted to nitrate concentration data using $D_{s_{NO_3^-}} = 160\ cm^2\ yr^{-1}$, $C_m = 0.040\ \mu mole\ cm^{-3}$, and $x_m = 5$ cm (Figure 7-8). The D_s value was obtained from the data of Li and Gregory (1974) for chloride diffusion (nitrate and chloride have about the same ionic mobility) in red clay at 0°C. The curve of best fit, shown in Figure 7-8, results in the value:

$$k_{NO_3^-} = 0.75\ yr^{-1}.$$

This value $k_{NO_3^-}$ can be contrasted with the value $k_{NO_3^-} = 158\ yr^{-1}$ obtained for nitrate reduction in near-shore sediments of the North Sea (Vanderborght et al., 1977b; see also Chapter 6). The much lower value found here reflects the much lower reactivity of organic matter in deep-sea sediments. This point was emphasized earlier when discussing organic matter decomposition via sulfate reduction, and it is apparent that lower reactivity is manifested in terms of lowered rates for other bacterial processes.

In the above calculation it was assumed that nitrate reduction in pelagic sediments could be described in terms of first order kinetics. However,

Froelich et al. present other profiles of dissolved nitrate concentration vs depth which are virtually straight lines, and in these cases a simple first order model, which predicts an exponential decrease with depth, cannot be correct. They suggest that the straight lines represent diffusion from the nitrate maximum, through a "neutral zone" where no nitrate reduction occurs, to a depth where denitrification is operative. In this way, first order kinetics may be retained. However, more data on deep-sea suboxic diagenesis is needed before this "neutral zone" concept can be accepted as a general phenomenon.

Diagenesis of Radioisotopes

Radioisotopes deposited in pelagic sediments, either by natural or artificial means, provide a very useful method for quantifying rate processes during early diagenesis. The most outstanding example is the use of radioactive decay to determine rates of sedimentation. This is most easily accomplished if bioturbation is absent, there is steady diagenesis, and the radioisotope is present in a highly insoluble form (to avoid diagenetic redistribution via pore water diffusion). These conditions are often met in pelagic sediments below the zone of bioturbation, which means below about 10 centimeters depth. The appropriate diagenetic equation in this case is very simple and considers only deposition and radioactive decay. Within the zone of bioturbation, radioisotopes are also useful in quantifying the rate of biodiffusion (see Chapter 3) if the rate of deposition is known independently, or if it is much slower than bioturbation.

Some radioisotopes undergo release to solution, adsorption, diffusion, and other processes, and thus require much more complex expressions to explain their distribution in sediments. Nevertheless, because the kinetics of radioactive decay (and production) are accurately known, it is possible to construct diagenetic models for such isotopes which rest on a reasonably firm basis as compared to many of the kinetic models discussed in this book which are based on rather crude assumptions as to rate laws and mechanisms. A good example is that of ^{226}Ra. This isotope is produced in sediments by the decay of ^{230}Th which, in turn, occurs in deep-sea sediments in excess over its parent ^{234}U. The ^{230}Th is present in excess because of its insolubility in seawater and preferential removal to the bottom relative to ^{234}U. The ^{226}Ra produced by decay of solid ^{230}Th is not restricted to solids and leaks into the surrounding pore water. This soluble ^{226}Ra is then free to diffuse and, as a result, it migrates upward and outward resulting in a deficiency of total (mainly solid) ^{226}Ra relative to ^{230}Th in the upper portions of the sediments.

A diagenetic model for the distribution of dissolved and adsorbed ^{226}Ra in pelagic red clay sediments of the north equatorial Pacific Ocean has been constructed by Cochran (1980) and tested by him against field measurements. Cochran's assumptions are:

1. Only a portion of the ^{226}Ra produced by the decay of solid ^{230}Th is ejected to solution with the remainder staying within the solid.
2. Radium-226 decays both in solution and adsorbed on particle surfaces via classical first order kinetics, with a half life of 1,622 years.
3. Adsorbed ^{226}Ra forms only by adsorption from solution and not via direct production from ^{230}Th. The adsorption follows a simple linear isotherm.
4. Pore water bioturbation is sufficiently slower than molecular diffusion that it can be neglected, i.e., $D_s \gg D_B$; $D_s \gg D_I$.
5. Particle bioturbation occurs within a fixed depth zone at the top of the sediment column and it can be described in terms of a constant bio-diffusion constant D_B within this zone of thickness L.
6. Compaction is negligible (justified by porosity measurements) over the depth zone of interest. Also there are no porosity gradients, externally impressed water flow, etc.
7. Steady state diagenesis is present.

With these assumptions we can use equation (4-52) of Chapter 4 to derive directly the diagenetic equation for dissolved ^{226}Ra. For the zone of bioturbation ($0 \leqq x \leqq L$), it is:

$$(D_s + KD_B)\frac{\partial^2 C}{\partial x^2} - (1 + K)\omega\frac{\partial C}{\partial x} - (1 + K)\lambda_{\mathrm{Ra}}C + P = 0. \quad (7\text{-}27)$$

(Here the reasonable assumption $D_s + D_B \approx D_s$ is made.) Below the zone of bioturbation the same equation applies but with $D_B = 0$. Here:

C = concentration of ^{226}Ra in atoms per unit volume of pore water;
λ_{Ra} = decay constant for ^{226}Ra;
P = rate of production of ^{226}Ra from ^{230}Th.

The rate of production, P, was given by the expressions;

In the zone of bioturbation ($0 \leqq x \leqq L$):

$$P = f\lambda_{\mathrm{Th}}FN^{L}_{\mathrm{Th}} + f'\lambda_{\mathrm{Th}}FN_{\mathrm{U}}. \quad (7\text{-}28)$$

Below the zone of bioturbation:

$$P = f\lambda_{Th}FN_{Th}^{L}\exp\left[-\frac{\lambda_{Th}}{\omega}(x - L)\right] + f'\lambda_{Th}FN_{U}, \qquad (7\text{-}29)$$

where N_{Th}^{L} = concentration of excess ^{230}Th (over that at radioactive equilibrium with ^{234}U) in the bioturbation zone in atoms per unit mass of total solids;

N_U = concentration of ^{234}U-supported ^{230}Th in atoms per unit mass of total solids;

f = fraction of decays of excess ^{230}Th which eject Ra atoms to solution;

f' = fraction of decays of ^{234}U-supported ^{230}Th which eject Ra atoms to solution;

$F = \bar{\rho}_s(1 - \phi)/\phi$;

λ_{Th} = decay constant of ^{230}Th.

The first terms on the right-hand side of each equation represent ^{226}Ra produced by the decay of excess ^{230}Th. Within the bioturbation zone excess ^{230}Th is homogeneously distributed because of its long half life (75,200 yr) relative to the rate of bioturbational mixing. Below the zone of bioturbation the production term decreases with depth as the excess ^{230}Th decays away. The second terms on the right-hand side represent ^{226}Ra produced by the decay of ^{230}Th at equilibrium with ^{234}U. Values of f and f' were derived by Cochran from various arguments based on laboratory emanation experiments and geometrical considerations.

Equation (7-27) with the appropriate substitution for P was solved separately for the zone of bioturbation and below it. The boundary conditions employed were:

$$x = 0,\ C = C_0 \text{ (value for overlying water)};$$
$$x = L,\ C_1 = C_2;$$
$$x = L,\ D_s(\partial C_2/\partial x) = (D_s + KD_B)(\partial C_1/\partial x);$$
$$x \to \infty,\ C \to (f'\lambda_{Th}FN_U/\lambda_{Ra}(1 + K).$$

Here C_1 and C_2 refer to concentrations within and below the zone of bioturbation respectively. The lower boundary conditions refer to radioactive equilibrium between dissolved-plus-adsorbed ^{226}Ra and both ^{230}Th and ^{234}U.

Cochran solved (7-27) for the given boundary conditions and plotted the results in terms of activity $A(=\lambda_{Ra}C)$ for a series of different K values until a best fit with measured pore water ^{226}Ra concentrations was obtained. To do this he used the estimated values $D_s = 95$ cm^2 yr^{-1}, $D_B = 0.032$ cm^2 yr^{-1}, $f = 0.6$, $f' = 0.3$, along with known or measured values of ω, A^L_{Th} $(=\lambda_{\mathrm{Th}} N^L_{\mathrm{Th}})$, A_{U} $(=\lambda_{\mathrm{Th}} N_{\mathrm{U}})$, F, L, λ_{Th}, and λ_{Ra}. Results for one core, shown in Figure 7-9, show a best fit for $K = 5{,}000$. This shows that Ra should be strongly adsorbed in this sediment. Independent measurement of ^{226}Ra given off by the particles due to desorption, assuming reversible equilibrium, gives a value of about 10,000. This discrepancy is probably due to a failure to reproduce the conditions of adsorption-desorption (e.g., pH) in the laboratory. At any rate, the agreement is good considering the many parameters involved whose values are not known accurately.

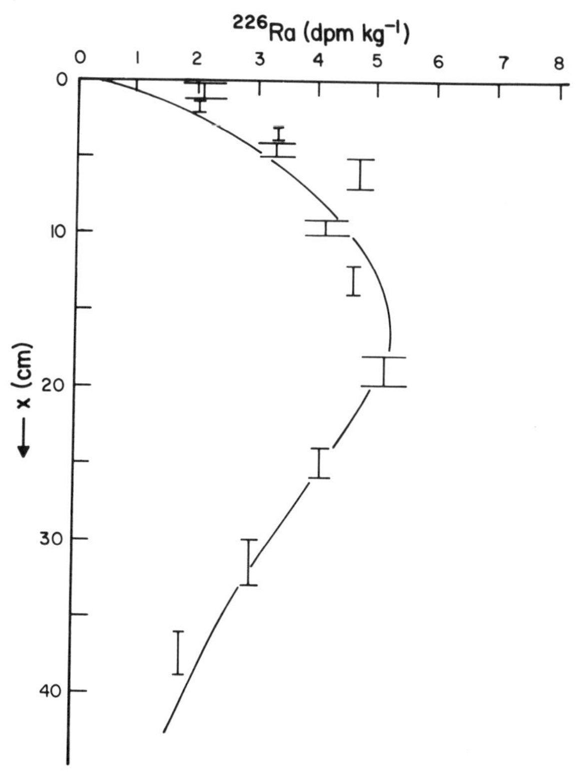

FIGURE 7-9. Plot of measured data for dissolved ^{226}Ra vs depth in a sediment from the Pacific Ocean. A theoretical curve, calculated according to the solution of equation (7-27), is plotted for the value $K = 5000$. (After Cochran, 1980.)

Volcanic-Seawater Reaction

Some pelagic sediments are notable for their high content of volcaniclastic debris. This is especially true of sediments deposited near sites of submarine volcanism, such as the mid-oceanic ridges. Studies of DSDP cores have often found deeply buried (several hundred meters depth) sediments close to basalt basement which are enriched in volcanic ash. By the theory of seafloor spreading, proximity to basement implies proximity to the ridge crest at the time of deposition, which would explain the enrichment in volcanics, especially volcanic glass and plagioclase (Peterson and Griffiin, 1964). This volcanic material readily reacts with seawater to form new silicate minerals, chiefly smectites and zeolites. Careful documentation of this reaction, in terms of diagenetic changes with depth in both solid and pore fluid composition, has been done by Perry et al. (1976). They showed that formation of smectite with increasing depth at the expense of volcanic material results in a decrease in dissolved Mg^{++} and an increase in dissolved Ca^{++} with depth. Further proof of authigenic mineral formation was obtained by means of oxygen isotopic analyses of both the smectite and interstitial water.

In other DSDP sediments changes in dissolved Mg^{++} and Ca^{++} in the pore waters have been interpreted in terms of similar reactions, and diagenetic models have been constructed to explain the profiles. Lerman (1975) chose to describe dissolved Mg^{++} and Ca^{++} in terms of a steady state diagenetic model whose basic assumptions are:

1. Mineral reactions involving Mg^{++} and Ca^{++} occur at all depths by means of linear rate laws. In other words, the rate of dissolution or precipitation is proportional to the first power of the degree of undersaturation or supersaturation ($n = 1$ in equation (5-42) of Chapter 5).
2. The effects of compaction on D_s, ω, and ϕ are all small and can be neglected.
3. Adsorption of Ca^{++} and Mg^{++} can be neglected.
4. Bioturbation is irrelevant since sediment depths of several hundred meters are involved.
5. Steady state diagenesis is attained.

Under these conditions, the diagenetic equation used by Lerman is:

$$D_s \frac{\partial^2 C}{\partial x^2} - \omega \frac{\partial C}{\partial x} + k_1(C_{eq_1} - C) - k_2(C - C_{eq_2}) = 0, \qquad (7\text{-}30)$$

where k_1, k_2 = rate constants for dissolving and precipitating phases respectively;

C_{eq1}, C_{eq2} = saturation concentration for the dissolving and precipitating phases respectively.

Substituting:

$$k = k_1 + k_2,$$
$$J = k_1 C_{eq_1} + k_2 C_{eq_2},$$

Lerman rewrote (7-30) as:

$$D_s \frac{\partial^2 C}{\partial x^2} - \omega \frac{\partial C}{\partial x} - kC + J = 0, \qquad (7\text{-}31)$$

where $J \neq f(x)$.

Lerman solved (7-31) for fixed concentrations at fixed upper and lower boundaries. From curve fitting plus the use of reasonable values of ω and D_s, he obtained values of k and J for Mg^{++} and Ca^{++}. He was thereby able to characterize profiles in terms of whether J was less than, equal to, or greater than zero for each ion. This was then used to deduce the type of mineral reaction involved, which included dissolution and precipitation of both carbonates and silicates.

McDuff and Gieskes (1976) have attacked Lerman's interpretation of the dissolved Ca^{++} and Mg^{++} data for DSDP sediments. They found, through the use of resistivity measurements, that the formation factor for molecular diffusion, F, increased with depth as a result of compaction. This means that the molecular diffusion coefficient for Ca^{++} and Mg^{++} should decrease with depth, and not remain constant as assumed by Lerman. By using a variable diffusion coefficient, based on their formation factor determinations, McDuff and Gieskes were able to explain the same Ca^{++} and Mg^{++} profiles studied by Lerman without invoking mineral dissolution or precipitation. The assumptions of their diagenetic model are:

1. No diagenetic chemical reactions occur between the fixed upper and lower boundaries.
2. Adsorption and bioturbation are negligible.
3. Externally impressed water flow is absent.
4. Steady state diagenesis, including steady state compaction, is present.

With these assumptions, the diagenetic equation from (3-74) is:

$$\frac{\partial\left(\phi D_s \frac{\partial C}{\partial x}\right)}{\partial x} - \frac{\partial(\phi v C)}{\partial x} = 0. \qquad (7\text{-}32)$$

Since there is steady state compaction, equation (7-32) can be simplified using equation (3-20) of Chapter 3:

$$\phi v = \phi_x \omega_x,$$

where the subscript x refers to a depth below which there is unappreciable compaction. Also, from equation (3-19) of Chapter 3:

$$\omega_x = \omega_0 \left(\frac{1 - \phi_0}{1 - \phi_x} \right),$$

where the subscript zero refers to the sediment-water interface. Substitution of (3-19) and (3-20) in (7-32) yields the simplified expression:

$$\frac{\partial \left(\phi D_s \frac{\partial C}{\partial x} \right)}{\partial x} - \omega_0 \phi_x \left(\frac{1 - \phi_0}{1 - \phi_x} \right) \frac{\partial C}{\partial x} = 0. \qquad (7\text{-}33)$$

The depth dependence of D_s was determined by measurements of the formation factor F on the same sediments using the relation:

$$D_s = \frac{D_s(0) F(x) \phi(x)}{F(0) \phi(0)} \qquad (7\text{-}34)$$

where (0) refers to the sediment-water interface. Values of F were found to obey the empirical relation

$$F = a\phi^{-n}. \qquad (7\text{-}35)$$

The values of a and n varied from one sediment to another and sometimes within different depth ranges of the same sediment. Once equations of the form (7-35) were established for each sediment section, then values of F could be determined for other sediment samples where F had not been measured, through the use of porosity measurements and equation (7-35). This made possible calculation of $D_s(x)$ for all depths.

Equation (7-33) was solved by numerical methods using measurements of porosity and F for fixed concentrations at the boundaries $x = 0$ (sediment-water interface) and $x = x_b$ (depth of bottommost sample taken). The concentration at $x = 0$ was chosen as that for the overlying seawater. At $x = x_b$, the concentration was chosen to "minimize the standard deviation of the observed points relative to the calculated profile" (McDuff and Gieskes, 1976, p. 5). The lower boundary condition, although lacking independent justification, is similar to that used by Lerman.

The numerical solutions to equation (7-33) fit very well the Ca^{++} and Mg^{++} data of several drill sites, as shown in Figure 7-10. (Mg^{++} data are omitted, but, because there is a linear correlation between Mg^{++} and Ca^{++}, they are equally well fitted by the theoretical curves.) This indicates that the concentration profiles of Ca^{++} and Mg^{++} are not *necessarily* due to diagenetic reactions occurring within the depth range of interest.

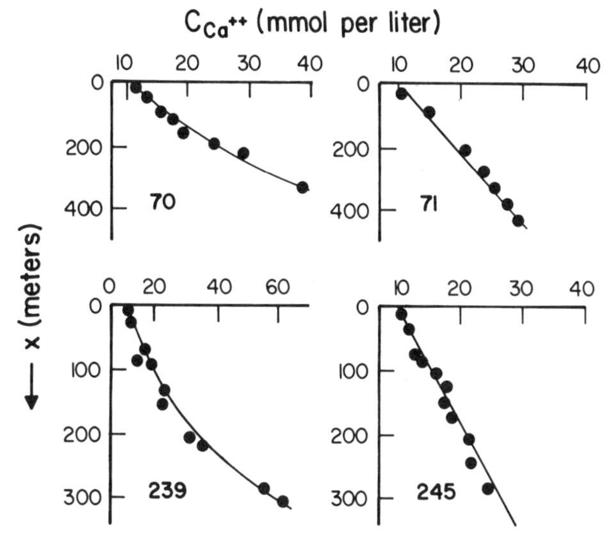

FIGURE 7-10. Dissolved Ca^{++} concentration data for DSDP pelagic sediment fitted to the advection-diffusion model of McDuff and Gieskes (1976). Numbers next to each curve denote DSDP drilling sites. Note that depth scale is in *meters*. (After McDuff and Gieskes, 1976.)

The model used by McDuff and Gieskes can equally well explain the curvature in the profiles as being due to decreases in D_s with depth. These authors thereby conclude that the profiles are caused by diagenetic reactions occurring only at depths *below* those sampled. In other words, diagenetic chemical reaction is present as a boundary condition. This is reasonable in the light of what was stated at the beginning of this section, that increased reaction betwen seawater Ca^{++} and Mg^{++} and minerals occurs in the lower portion of the DSDP sediment column where greater concentrations of reactive volcanic debris are found.

8

Non-Marine Sediments

The category of non-marine sediments is used here broadly to include all sediments deposited in waters whose chemical composition is distinctly different from that of seawater. For our purposes this includes lakes and rivers, supersaline lagoons, and some estuaries where salinities, due to mixing of river water with seawater, are distinctly lower than that found in the open ocean. Within this category we distinguish fresh and brackish sediments (less saline than seawater) from supersaline sediments (more saline than seawater). Most discussion here will be restricted to fresh and brackish sediments because they have been more extensively studied from an early diagenetic (modeling) standpoint.

Fresh and Brackish Water Sediments

The interstitial waters of fresh and brackish sediments, in contrast to marine sediments, are distinctive in both their chemical composition and total salt content (salinity). Not only do they vary from one lake or estuary to another, reflecting differences in overlying water chemistry, but they may also show distinct variations with depth within a single sediment core. These variations reflect both chemical diagenetic changes, and historical or short-term changes in the composition and salinity of the overlying water. In this section we will be concerned with three aspects of early diagenesis in low-salinity non-marine sediments. They are: (1) diffusion of salt (chloride) in sediments as a result of salinity fluctuation in the overlying water, (2) diagenetic changes accompanying authigenic iron mineral formation, and (3) the diagenetic redistribution of manganese.

Salinity Fluctuations

Salinity fluctuations in sediments respond, via diffusive exchange, to corresponding fluctuations in the overlying water. This gives rise, by definition, to non-steady state diagenesis. In brackish estuaries the fluctuating salinities are caused by seasonal and longer term variations in river flow which results in varying amounts of fresh water being mixed with normal seawater. If one were to focus on chloride ion in an estuarine sediment pore water, changes due solely to salinity fluctuation in the overlying water could be followed with depth in the sediment. This is possible because Cl^- is an inert tracer which means that it does not undergo any diagenetic chemical reactions nor is it adsorbed to any appreciable extent. In this case the appropriate diagenetic equation expressing the effects of bioturbation, molecular diffusion, pore water flow, and burial from equation (3–74) is:

$$\frac{\partial(\phi C)}{\partial t} = \frac{\partial\left[D_B \frac{\partial(\phi C)}{\partial x} + \phi(D_s + D_I)\frac{\partial C}{\partial x}\right]}{\partial x} - \frac{\partial(\phi v C)}{\partial x}, \qquad (8\text{-}1)$$

where C here represents the concentration of chloride ion.

To simplify our discussion we will assume that there is no compactive or other type of water flow and that the sediment solids are undergoing a constant rate of deposition. In this case, $\partial\phi/\partial x = 0$, and $v = \omega =$ constant. If in addition D_B and D_I are constant within the zone of bioturbation and zero below it, and if D_s is a constant at all depths, we have:

$$\frac{\partial C}{\partial t} = D_T \frac{\partial^2 C}{\partial x^2} - \omega \frac{\partial C}{\partial x}, \qquad (8\text{-}2)$$

where $0 \leqq x \leqq L: D_T = D_B + D_I + D_s;\ C = C_1;$

$x > L: D_T = D_s;\ C = C_2;$

(L = thickness of bioturbation zone).

Here the subscripts 1 and 2 refer to sediments within the bioturbation zone and below it, respectively.

Solution of equation (8–2) for C_1 and C_2 depends upon the boundary conditions. The lower boundary condition is that, because of bioturbational and diffusive mixing, a constant chloride concentration is approached with depth, in other words as $x \to \infty$, $\partial C_2/\partial x = 0$. At the depth L we have the continuity expressions:

$$C_{1_L} = C_{2_L}, \qquad (8\text{-}3)$$

$$(D_B + D_I + D_S)\left(\frac{\partial C_1}{\partial x}\right)_L = D_S\left(\frac{\partial C_2}{\partial x}\right)_L \qquad (8\text{-}4)$$

The upper boundary condition is that at $x = 0$, the concentration is the same as in the overlying water. The concentration in the overlying water, in turn, for brackish estuaries, is a function of time. For example, in the brackish northern portion of Chesapeake Bay (U.S.A), chloride concentration varies yearly due to seasonal flow from the Susquehanna River, and over longer time scales due to climatic fluctuations. Holdren et al. (1975) have suggested use of the following periodic function for chloride concentrations in northern Chesapeake Bay:

$$C = C_0 + A\cos(2\pi\lambda_1 t) + B\cos(2\pi\lambda_2 t), \tag{8-5}$$

where C = concentration of chloride in overlying water;
C_0 = long-term mean chloride concentration;
$A; B$ = empirical constant;
λ_1 = seasonal periodicity (one year^{-1});
λ_2 = long-term periodicity.

Matisoff (1980) has suggested that the long-term change is better represented as a linear function of time for the past 10 years. However, for our purposes the approximation of Holdren et al. is sufficient since it leads to practically the same results as the calculations of Matisoff.

Equation (8-2) with the upper boundary condition (8-5) has been solved by Holdren et al. (1975) to describe chloride concentrations in interstitial waters of sediments from a station in northern Chesapeake Bay. However, these authors (and also Matisoff, 1980) assume a constant value of D_T at all depths, which is tantamount to neglecting bioturbation. In other words, concentrations C_1 and C_2 are not distinguished from one another. The solution of Holdren et al. (1975) is:

$$C = C_0 + A\cos\left[\lambda_1 t - a_1^{1/2}\sin\left(\frac{\alpha_1}{2}\right)x\right]\exp\left[\frac{\omega}{2D_T}x - a_1^{1/2}\cos\left(\frac{\alpha_1}{2}\right)x\right] + B\cos\left[\lambda_2 t - a_2^{1/2}\sin\left(\frac{\alpha_2}{2}\right)x\right]\exp\left[\frac{\omega}{2D_T}x - a_2^{1/2}\cos\left(\frac{\alpha_2}{2}\right)x\right], \tag{8-6}$$

where $a = \left(\frac{\omega^4}{16D_T^4} + \frac{\lambda^2}{D_T^2}\right);$

$\alpha = \tan^{-1}(4D_T\lambda/\omega^2).$

For estimated values of ω and λ_2, various values of D_T were tried until a best fit to the measured data was obtained. Results (see Figure 8-1) were shown to be relatively insensitive to ω and λ_2, but quite sensitive to the value chosen for D_T. The best fit D_T value was $D_T = 5 \times 10^{-6}$ cm^2 sec^{-1}. This is practically the same as that which would be expected for molecular diffusion of chloride ion in these sediments (Li and Gregory, 1974). Thus, the neglect of bioturbation appears to be justified. Nevertheless, a more exact treatment of the problem should involve consideration of enhanced diffusion near the sediment-water interface due to bioturbation as well as to variations in D_T with time due to seasonal temperature fluctuations.

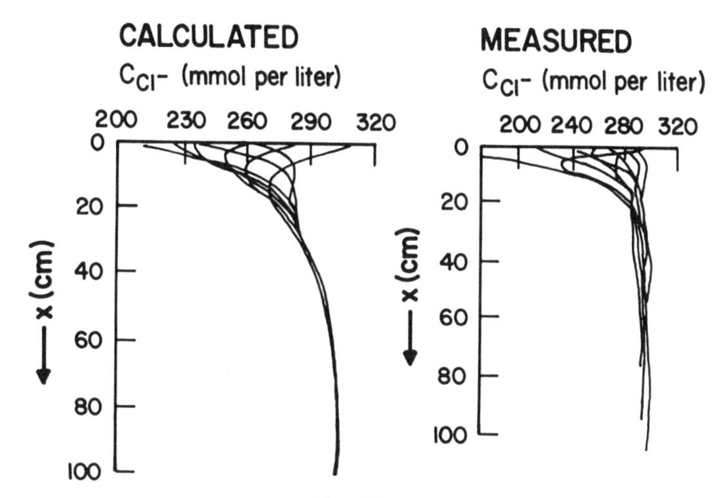

From Marine Chemistry in the Coastal Environment

FIGURE 8-1. Calculated and measured dissolved chloride profiles for a sediment from northern Chesapeake Bay, USA. Each curve represents a different time of year. (After Holdren et al., 1975.)

The fact that several points near the sediment-water interface were not well fitted by theoretical curves, in both the studies of Holdren et al. and Matisoff suggests a need for refinement of their models.

The interesting overall result of the above studies of salinity (as represented by chloride concentration) in estuarine sediment pore waters is the wave pattern of the concentration-vs-depth curves. Their change with time is reminiscent of temperature profiles found in soils and sediments. Each curve represents seasonal concentration variations which are severely damped with depth due to diffusion. The distance of dampening corresponds to the average distance traveled by a chloride ion in one year which, using the relation $x^2 = 2D_s t$, with $D_s = 200\ \text{cm}^2\ \text{yr}^{-1}$ (Tables 3-2 and 3-3), is 20 centimeters (see Figure 8-1). The asymptotic value approached at depth represents a mean annual salinity which decreases upwards due to long-term climatic fluctuations. The extent of long-term fluctuations, like seasonal fluctuations, however, must also undergo diffusional damping given enough time. *In general, molecular diffusion is an effective process for altering and even obliterating the record of historical changes as it is recorded in the composition of interstitial water.*

What has been said above for estuarine sediments also applies to many freshwater lakes. Concentration variations in lakes due to changes in river input, rainfall, or evaporation should be detectable in sediment pore waters. For example, Lerman and Weiler (1970) have documented increases in Cl^- and Na^+ in the pore water of sediments from Lake Ontario, due largely to increases of these ions in the overlying water. The increases are attributed to a rising inflow to the lake over the past century of NaCl contributed by industrial, agricultural, municipal, and mining wastes. Using a linear increase in lake water concentration of both ions over the past 70 years as an upper boundary condition, Lerman and Weiler solved equation (8-2) (for constant D_T) and, using reasonable values for D_T, were able to fit theoretical curves to the measured pore water data.

Iron Diagenesis

Besides variable salinity, another distinctive aspect of low-salinity non-marine sediments is the behavior of iron during early diagenesis. In the interstitial waters of anoxic, organic-rich, non-marine sediments one commonly encounters high concentrations of dissolved iron and the resultant formation, during early diagenesis, of reduced authigenic iron minerals. These minerals are of special interest since several of them (e.g., vivianite $Fe_3(PO_4)_2 \cdot 8H_2O$; greigite Fe_3S_4; siderite, $FeCO_3$) are otherwise rare in normal marine sediments. Thus, their occurrence in ancient shales may serve as useful paleosalinity indicators (Berner, 1971). In the marine environment, as shown in the previous chapter, the authigenic iron mineral that forms during the early diagenesis of organic-rich, anoxic sediments is normally pyrite, FeS_2. There is sufficient dissolved sulfate in seawater to provide enough H_2S to convert almost all reactive detrital iron to FeS_2. However, in brackish and fresh water sediments, because of low salinity and, therefore, low initial SO_4^{--} concentrations, sulfate is normally totally reduced at very shallow levels of the sediment. Not enough H_2S is produced to "titrate" the available reactive iron. As a result, once sulfate is depleted and, all remaining H_2S is precipitated to form highly insoluble FeS minerals, the concentration of dissolved Fe^{++} can build up by continued bacterial reduction of iron oxides. This continues until saturation with vivianite, siderite, or other iron minerals is sufficiently exceeded that precipitation occurs. Meanwhile the FeS minerals (greigite, mackinawite) are not converted to pyrite, as they are in marine sediments, because of a lack of H_2S. Consequently, upon burial they persist and, together with vivianite and/or siderite, constitute a diagnostic authigenic assemblage.

What has been said here does not apply to freshwater sediments *unusually* high in organic matter, in other words to swamps. Here the organic matter itself can furnish sufficient sulfur that liberation of it as H_2S during diagenetic bacterial degradation allows conversion of most reactive detrital iron to pyrite (or marcasite). In this way the common association of pyrite and marcasite with coal may be explained.

Formation of authigenic vivianite in an anoxic lake sediment (Lake Greifensee, Switzerland) has been well documented by Emerson and Widmer (1978). Here sulfate is exhausted in the top ~2 cm of sediment so that there is no sulfide production below this depth. Organic matter decomposition by other kinds of bacteria continues and results in the liberation to solution of phosphorus from organic compounds and iron from detrital iron minerals. Vivianite is supersaturated from the sediment-water interface down to about 20 cm where equilibrium is attained. The pore water is just saturated with respect to $FeS\text{-}Fe_3S_4$ minerals at all depths, and continual increases in dissolved Fe^{++} with depth are matched by corresponding drops in dissolved sulfide. Total dissolved phosphate drops off continually with depth due to precipitation, reaching a constant low value at about 20 cm. Data for Fe^{++} and PO_4^{-3} are shown in Figure 8-2. Also, the sediment is finely varved, indicating minimal bioturbation. Values of mass sedimentation rate and ω derived from varve counts are shown vs depth in Figure 8-3. Constant mass sedimentation rate suggests

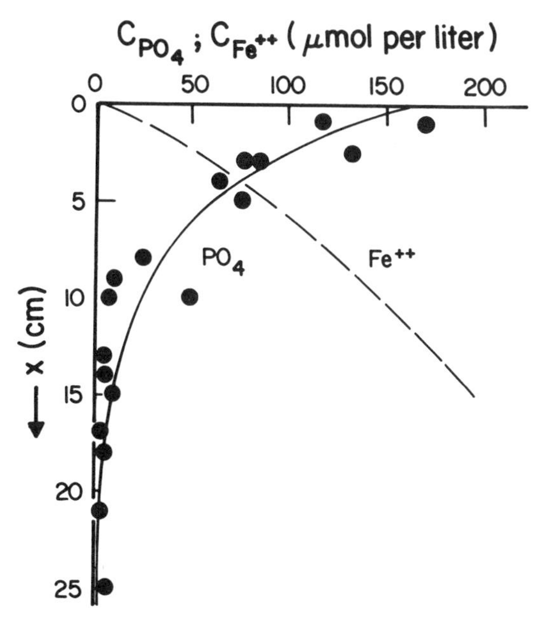

FIGURE 8-2. Plot of dissolved phosphate vs depth for sediments of Lake Greifensee, Switzerland. Shown also is the general distribution of dissolved Fe^{++}. The phosphate data is fitted by the empirical expression (8-14) given in the text. (Adapted from Emerson and Widmer, 1978.)

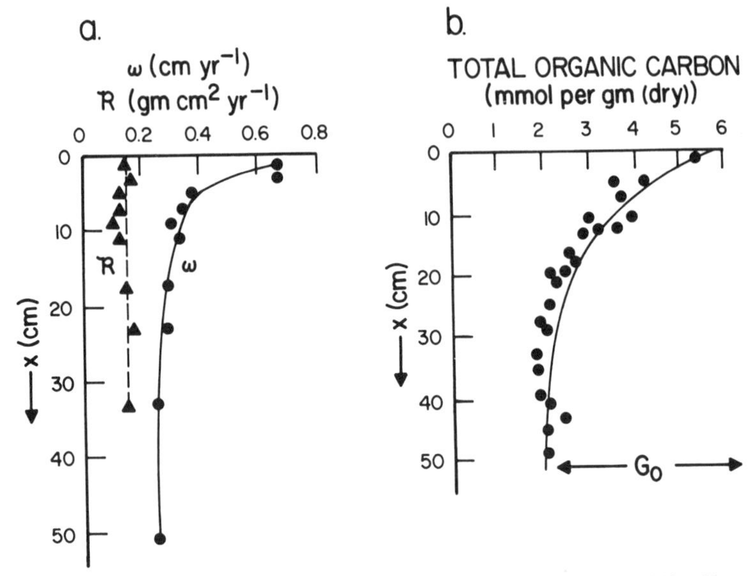

FIGURE 8-3. Early diagenesis in sediments of Lake Greifensee, Switzerland. (After Emerson and Widmer, 1978).

(a) Rate of deposition, as deduced from varve counts, vs depth. Data are expressed both as ω (cm yr^{-1}) and as mass flux, $\mathscr{R}$(gm cm^{-2} yr^{-1}).

(b) Organic carbon concentration (dry weight) vs depth. Data are fitted by the theoretical expression (8-9) of text.

steady state diagenesis (including compaction), which is in agreement with the mathematical calculations of Emerson and Widmer that indicate that steady state for phosphate is a good assumption.

These data, along with diagenetic modeling, can be used to deduce the rate-controlling mechanism of vivianite precipitation in the Lake Greifensee sediments. The reasoning goes as follows. For the situation described above, the appropriate diagenetic equation from equations (3-20), (3-74), and (4-52) for dissolved phosphate is:

$$(1 + K_P)\frac{\partial C}{\partial t} = \frac{1}{\phi}\frac{\partial\left(\phi D_s \dfrac{\partial C}{\partial x}\right)}{\partial x} - \left[\frac{(1 + K_P)\phi_x\omega_x}{\phi}\right]\frac{\partial C}{\partial x} + R_{biol} - R_{pptn} = 0, \tag{8-7}$$

where C = concentration of dissolved phosphate. To simplify calculation we may assume that a constant average value of ϕ over the top 20 cm, $\bar{\phi}$, can be substituted for ϕ in equation (8-7), and that D_s is roughly constant with depth. These sorts of assumptions for a similar situation in a marine varved sediment (Murray et al., 1978) have been shown to be justifiable for the level of accuracy we seek. Thus, equation (8-7) can be reduced to:

$$D_s \frac{\partial^2 C}{\partial x^2} - (1 + K)\bar{\omega}\frac{\partial C}{\partial x} + R_{biol} - R_{pptn} = 0, \tag{8-8}$$

where $\bar{\omega} = \phi_x \omega_x / \bar{\phi}$.

Evaluation of the R_{biol} term is possible from Emerson and Widmer's plot of organic carbon vs depth (Figure 8-3). Although they state their belief that the organic carbon gradient mainly reflects historic increases in carbon deposition rate, one can also interpret the C-vs-depth curve solely in terms of diagenetic bacterial decomposition with depth. If possible historic changes are kept in mind, then our assumption of a purely diagenetic carbon change means that our calculated values of R_{biol} will be a maximum at each depth. Assuming steady state diagenesis, and using the one-G model for microbial decomposition (equation 4-73):

$$G = G_0 \exp\left[(-k/\bar{\omega})x\right], \tag{8-9}$$

$$R_{biol} = -\alpha_P \bar{F}\frac{\partial G}{\partial t_{biol}}, \tag{8-10}$$

$$R_{biol} = \alpha_P \bar{F} k G_0 \exp\left[(-k/\bar{\omega})x\right], \tag{8-11}$$

where $\alpha_p = P{:}C$ mole ratio of decomposing organic matter;
$\bar{F} = (1 - \bar{\phi})\bar{\rho}_s/\bar{\phi}$.

Substitution of (8-11) in (8-8) and solving for R_{pptn} yields:

$$R_{pptn} = D_s \frac{\partial^2 C}{\partial x^2} - (1 + K_P)\bar{\omega}\frac{\partial C}{\partial x} + \alpha_P \bar{F} k G_0 \exp\left[(-k/\bar{\omega})x\right]. \tag{8-12}$$

We may identify G_0 as the difference between the concentration of total organic carbon at $x = 0$ and the asymptotic concentration where $x \to \infty$. Also, fitting of (8-9) to the curve for organic carbon of Figure 8-3 yields $(k/\bar{\omega})$. From Figure 8-3 we obtain:

$$G_0 = 4{,}000\ \mu\text{mol gm}^{-1},$$
$$k/\bar{\omega} = 0.086\ \text{cm}^{-1}.$$

Also, we have from Emerson and Widmer's data:

$$\bar{\phi} = 0.88,$$
$$\bar{\omega} = 0.20\ \text{cm yr}^{-1},$$
$$\alpha_P = 0.007,$$
$$\bar{\rho}_s = 2.5\ \text{gm cm}^{-3}.$$

And we can assume the reasonable values (see Chapters 3 and 6):

$$D_s = 120\ \text{cm}^2\ \text{yr}^{-1},$$
$$K_P = 2.$$

Substituting these values, equation (8-12) becomes:

$$R_{pptn} = 120\frac{\partial^2 C}{\partial x^2} - 0.70\frac{\partial C}{\partial x} + 0.164 \exp(-0.086x) \quad (8\text{-}13)$$

where R_{pptn} is expressed in μmoles per cm^3 per year.

Now from the data of Figure 8-2, we can graphically obtain $\partial C/\partial x$ and $\partial^2 C/\partial x^2$ at each depth. However, to simplify matters I have assumed that phosphate decreases exponentially with depth and have fitted to the data of Figure 8-2 the empirical expression:

$$C = 0.16 \exp(-0.20x) + 0.002, \quad (8\text{-}14)$$

where C is in μmole per cm^3. Taking the first and second derivatives of this expression and substituting the results in (8-13) we obtain:

$$R_{pptn} = 0.790 \exp(-0.20x) + 0.164 \exp(-0.086x). \quad (8\text{-}15)$$

From (8-15) we can deduce the rate-controlling mechanism for vivianite precipitation. Emerson and Widmer have calculated via an expression analogous to equation (5-33) of Chapter 5 the supersaturation $(C - C_{eq})$ at each depth where C_{eq} represents equilibrium with vivianite at the vivianite surface. At $x = 4$ cm, maximal supersaturation values for most cores are found. Combining the supersaturation (0.04 μmole cm^{-3}) at $x = 4$ with R_{pptn} at this depth, calculated from equation (8-15), we obtain:

$$\frac{R_{pptn}}{(C - C_{eq})} = 12 \text{ yr}^{-1}. \quad (8\text{-}16)$$

For diffusion-controlled precipitation we have:

$$\frac{R_{pptn}}{(C - C_{eq})} = \frac{\bar{A}_{viv}\phi D_s}{r_c}. \quad (8\text{-}17)$$

Emerson and Widmer, using an electron microprobe, found that particles of vivianite average about 10 μm in diameter. From this (assuming spherical grains) and the relation $\bar{A}_{viv} = 3N(1 - \phi)/\phi r_c$ (where N = volume fraction of solids that is vivianite) we obtain:

$$r_c = 5 \times 10^{-4} \text{ cm},$$
$$\bar{A}_{viv} \approx 0.6 \text{ cm}^2 \text{ cm}^{-3} \text{ of pore water.}$$

Substituting these values, along with $\phi = 0.88$ and $D_s = 120$ cm^2/yr, in (8-17), we calculate for diffusion control:

$$\frac{R_{pptn}}{(C - C_{eq})} = 1.3 \times 10^5 \text{ yr}^{-1}. \quad (8\text{-}18)$$

This value is about four orders of magnitude higher than that calculated for in situ vivianite growth from the diagenetic equation (8-16). Also, the in situ value is maximal, because of the assumption that organic carbon decreases with depth are purely diagenetic and not historical. Thus, it is apparent that vivianite precipitation is far slower than that predicted for control by molecular diffusion. This agrees with the deduction of Emerson and Widmer (1978). Surface chemical reaction must be the rate-controlling process for authigenic vivianite precipitation in this sediment and, very likely, in most other sediments where vivianite is forming.

Manganese Diagenesis

In addition to iron, dissolved manganese commonly builds up in the interstitial waters of organic-rich, brackish, and freshwater sediments. This comes about from the reduction of MnO_2 accompanying the bacterial decomposition of organic matter. Dissolved Mn^{++} commonly exceeds saturation with respect to various reduced manganese minerals, and as a result new authigenic phases may precipitate. The most common one is rhodochrosite ($MnCO_3$) but reddingite ($Mn_3(PO_4)_2 \cdot 3H_2O$) and various forms of MnS (Suess, 1979; Aller, 1977; Nriagu and Dell, 1974) in rarer situations may also form. Two sedimentary situations will be discussed here: that in Lake Michigan (Robbins and Callender, 1975) and that in Chesapeake Bay (Holdren et al., 1975; Holdren, 1977).

Robbins and Callender (1975) have developed a diagenetic model to explain the distribution of manganese, both dissolved and solid, in sediments of Lake Michigan. To check their model they obtained sedimentation rates (via ^{210}Pb) and chemical analyses of acid-soluble solid Mn and of dissolved Mn^{++} in the interstitial water. Qualitatively the diagenetic model of Michard (1971) for manganese was employed. At the sediment-water interface dissolved Mn^{++} diffuses into the overlying water which, due to mixing and precipitation of MnO_2 by dissolved O_2, maintains a low Mn^{++} concentration (such precipitation commonly results in the formation of Mn-nodules and coatings on rocks). From $x = 0$ to $x = 4$ cm, a neutral zone exists where neither oxidation of Mn^{++} nor reduction of MnO_2 occurs. In other words, only diffusion occurs within this zone. At depths from 4 to 8 cm reduction of MnO_2 occurs with the production of dissolved Mn^{++} in the pore water. Below 8 cm there is no dissolution and, instead, Mn^{++} is precipitated to form authigenic rhodochrosite, $MnCO_3$. Diagrammatic illustration of these zones is shown in Figure 8-4a.

Robbins and Callender, in their model, ignore bioturbation and equilibrium adsorption. Also, they assume steady state with respect to both Mn^{++} and porosity. Under these conditions the appropriate diagenetic equation for dissolved Mn^{++} from (3-20) and (3-74) is:

$$\phi D_s \frac{\partial^2 C}{\partial x^2} + \left[\frac{\partial(\phi D_s)}{\partial x} - \phi_x \omega_x\right]\frac{\partial C}{\partial x} + \phi R_{biol} + \phi R_{pptn} = 0. \quad (8\text{-}19)$$

Here R_{biol} represents reduction of MnO_2 to Mn^{++}, and R_{pptn} represents the precipitation of $MnCO_3$. Using the relation, $D_s = D_{so}\phi^2$ (Manheim, 1970) and known values of $\phi(x)$, ϕ_x, ω_x, and D_{so}, they evaluated the advective term within brackets and found that it varied only by a factor of two over the depth range of interest. To simplify matters they assumed an average value for this term of -0.2 cm per year. Also, the precipitation of $MnCO_3$ was assumed to follow the simple relation:

$$R_{pptn} = k_1(C - C_{eq}), \quad (8\text{-}20)$$

where C_{eq} represents saturation with respect to rhodochrosite.

The form of the biological production term was deduced from measurements of (acid-soluble) solid Mn with depth. The measured data are shown in Figure 8-4. From this data one can deduce the form of the expression

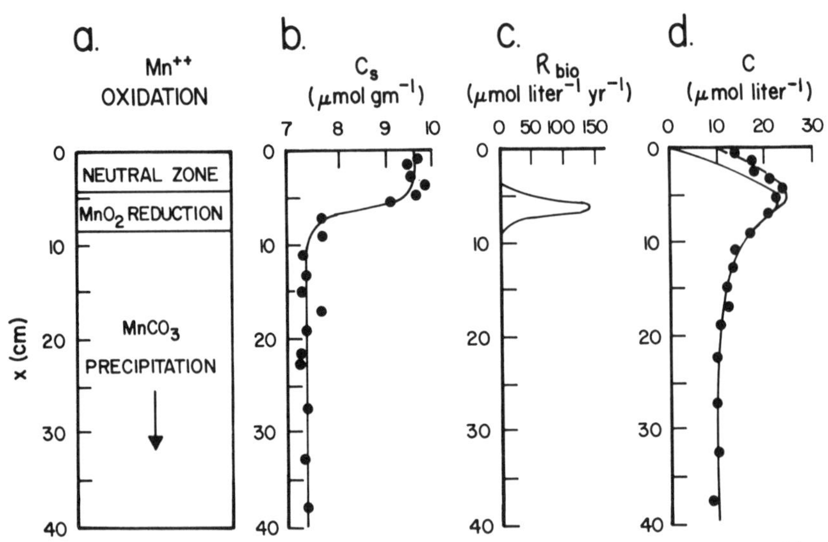

FIGURE 8-4. Diagenesis of manganese in Lake Michigan sediments. (Modified after Robbins and Callender, 1975.)
(a) Distribution of zones according to the model of Michard (1971).
(b) Distribution of acid-soluble solid manganese.
(c) Biological production term derived from the data shown in (b) (see equation 8-23).
(d) Dissolved Mn^{++} fitted by the theoretical curve calculated according to equation (8-24). The dashed line represents a better fit to the observed data (see text).

for solid Mn according to equation (3-73). In other words, assuming steady state and no bioturbation:

$$\frac{-\partial[(1-\phi)\bar{\rho}_s\omega C_s]}{\partial x} - \phi R_{biol} = 0. \tag{8-21}$$

If ω and ϕ are assumed to vary little over the depth range of interest (0–40 cm), upon substituting the relation $F = (1-\phi)\bar{\rho}_s/\phi$ into (8-21) we obtain:

$$R_{biol} = -\bar{\omega}\bar{F}\frac{\partial C_s}{\partial x}, \tag{8-22}$$

where $\bar{\omega}$, $\bar{F}$ = average values of ω and F over the depth range of interest (0–40 cm).

Values of $\partial C_s/\partial x$ were obtained by fitting a curve to the data for solid *Mn*, shown in Figure 8-4b, whose first derivative is a simple Gaussian function:

$$-\frac{\partial C_s}{\partial x} = \left[\frac{(C_{s_0} - C_{s_\infty})}{\sigma\pi^{1/2}}\right]\exp\{-[(x-x_m)/\sigma]^2\}. \tag{8-23}$$

Here the subscripts 0 and ∞ represent $x = 0$, and $x \to \infty$, respectively, and x_m = midpoint (mean) of the Gaussian. A least squares computer program was used to find the best values of C_{s_0}, C_{s_∞}, x_m, and σ. The results for R_{biol} are shown in Figure 8-4c. Note that by choosing appropriate values of σ, one obtains essentially no biological production of Mn^{++} either above 4 cm or below 8 cm depth. This is required by the zonal model.

Substitution of (8-20), (8-22), and (8-23) in (8-19), along with the various simplifying assumptions, leads to the final expression for dissolved Mn^{++}:

$$\bar{\phi}D_s\frac{\partial^2 C}{\partial x^2} + \omega^*\frac{\partial C}{\partial x} + \bar{\omega}\bar{F}\left(\frac{C_{s_0} - C_{s_\infty}}{\sigma\pi^{1/2}}\right)\exp\{-[(x-x_m)/\sigma]^2\}$$
$$-k_1(C - C_{eq}) = 0, \tag{8-24}$$

where

$$\omega^* = \left[\frac{\partial(\phi D_s)}{\partial x} - \phi_x\omega_x\right] = -0.2 \text{ cm yr}^{-1}.$$

Solution of equation (8-24) for the boundary conditions:

$$x = 0; \qquad C = 0,$$
$$x \to \infty; \qquad C \to C_{eq},$$

was done, which resulted in a complex expression. By use of a computer this expression was then fit to the pore water data for Mn^{++}, and the result is shown in Figure 8-4d. From the curve of best fit Robbins and Callender calculated the values:

$$D_s = 0.9 \times 10^{-6} \text{ cm}^2 \text{ sec}^{-1},$$
$$k_1 = 1 \text{ yr}^{-1},$$

which are good to $\pm 20\%$ in the sense that a less than 20% change in either parameter leads to curves indistinguishable within measurement error from that shown. The value of D_s is lower by about a factor of three than that expected for Mn^{++} in this sediment (see Table 3-4). Since the advective term is small relative to the other terms, this low value cannot be explained by means of retardation by equilibrium adsorption. (Equation (4-53) shows that if advection is negligible, at steady state the effects of adsorption cancel.) However, considering the many approximations and assumptions (especially of the simple first order dependency of $MnCO_3$ precipitation), the value obtained for D_s is reasonable.

A better fit to the interstitial water data is obtained if the upper boundary condition is changed to $x = 0$, $C = 0.011$ μmole cm^{-3}. This is shown in Figure 8-4d by the dashed line. This may mean that the value for the overlying water ($C \approx 0$) is not reached until slightly above the "sediment-water interface." Finding of a thin ($\sim$1 cm) layer of loose, flocculent material just above the sediment surface which was lost on coring and sampling suggests that the drop in concentration from 0.011 μmole cm^{-3} to zero may occur within this flocculent layer.

The asymptotic concentration at depth ($C = 10$ μM) is approximately the same as that calculated by Robbins and Callender for equilibrium with rhodochrosite. This bolsters their argument that rhodochrosite precipitation is the cause of the drop in concentration at depth. Also, the rate constant, k_1, can be compared to that expected for diffusion-controlled precipitation, in order to deduce the rate-controlling mechanism of precipitation. For diffusion-controlled precipitation one would expect:

$$k_1 = \frac{\bar{\phi}\bar{A}D_s}{r_c}. \tag{8-25}$$

Here $\bar{\phi} = 0.8$, $D_s \approx 10^{-6}$ cm^2/sec. Rhodochrosite itself was not studied in the sediment, but by analogy with vivianite discussed earlier, we can, for the purposes of calculation, assume that $r \approx 5 \times 10^{-4}$ cm. If so, $\bar{A} =$ 0.6 $\text{cm}^2{}_{\text{rhod}}$ cm^{-3} pore water. From these values:

$$k_1 = 30{,}000 \text{ yr}^{-1}.$$

This value is much bigger than that calculated via the model (1 yr^{-1}), and no reasonable adjustment of particle size can bring the two values into agreement. Therefore, as in the case of vivianite, the rate of authigenic rhodochrosite precipitation in freshwater lake sediments must be controlled by surface-chemical reactions and not by transport of ions to the rhodochrosite surface.

The formation of dissolved Mn^{++} and authigenic rhodochrosite has also been documented and modeled theoretically in brackish estuarine sediments. In the northern, least saline, portions of Chesapeake Bay (U.S.A.), Holdren (1977) has found very high concentrations of dissolved Mn^{++} in the pore waters of anoxic sediments, generally in the range 50–500 μmole per liter. (The maximum value for the Lake Michigan sediments discussed above is about 20 μmole per liter.) These high concentrations, however, were found only during the warmer portions of the year. By contrast, during the winter months concentrations at the same sampling sites ranged from only about 10 to 50 μmoles per liter. Holdren explains this seasonal fluctuation in terms of the effect of temperature on rates of bacterial activity. In the warmer months bacterial activity is higher and rates of Mn^{++} production accompanying organic matter decomposition are correspondingly higher. In addition, and most important to Holdren's diagenetic model (discussed below), increased bacterial activity produces higher concentrations of dissolved organic matter (DOM) in the pore water. This DOM is believed to inhibit the precipitation of authigenic rhodochrosite by interfering with its nucleation and growth. In other words, it acts as a poison. During the summer the large build-up in dissolved Mn^{++} causes high degrees of rhodochrosite supersaturation to be attained, but precipitation is so slow, due to organic inhibition, that saturation equilibrium is not attained at any depth. In the winter, by contrast, less Mn^{++} is produced, and precipitation is faster because less DOM is present. As a result, the pore waters appear to be at equilibrium with rhodochrosite. The level of supersaturation ($\Omega \approx 17$) attained in summer is relatively constant for all sediments and depths, so that Holdren adopts the convention of treating this state as being at (pseudo) equilibrium. In other words, a degree of supersaturation is attained beyond which appreciable precipitation may occur but below which there is very little precipitation, due to inhibition. This situation is reminiscent of the very high order kinetics found for the dissolution of calcite in seawater.

With these ideas in mind, Holdren adopts the following diagenetic model. The sediment is divided into two depth zones. From the sediment-water interface down to the maximum in Mn^{++} concentration (see Figure 8-5), no precipitation of rhodochrosite occurs and the dissolved Mn^{++} distribution is explained solely in terms of diffusion, depositional burial, and the reduction of MnO_2, which is assumed to follow first order kinetics. He ignores compaction, bioturbation, and equilibrium adsorption. Also, he assumes rapid re-attainment of steady state so that summer profiles can be described in terms of steady state diagenesis. Under these conditions the appropriate diagenetic equations are:

For dissolved Mn^{++} (C):

$$D_s \frac{\partial^2 C}{\partial x^2} - \omega \frac{\partial C}{\partial x} + kC_s = 0. \tag{8-26}$$

For solid $MnO_2(C_s)$:

$$-\omega \frac{\partial C_s}{\partial x} - kC_s = 0. \tag{8-27}$$

Solution for the boundary conditions: $x = 0$, $C = 0$, $C_s = C_{s_0}$ and $x \rightarrow \infty$, $\partial C/\partial x \rightarrow 0$, $C_s \rightarrow 0$, yields:

$$C = \left[\frac{\omega^2 C_{s_0}}{\omega^2 + kD}\right]\{1 - \exp[(-k/\omega)x]\}. \tag{8-28}$$

Below the maximum in Mn^{++} concentration, Holdren employs the concept of equilibrium reaction (see Chapter 4). He assumes that at all depths the concentration of Mn^{++} is at equilibrium (really pseudo-equilibrium—see above) with $MnCO_3$. The reaction employed is:

$$MnCO_3 + H^+ \rightleftharpoons Mn^{++} + HCO_3^-,$$

$$K^* = \frac{a_{Mn^{++}} a_{HCO_3^-}}{a_{H^+}}, \tag{8-29}$$

where K^* refers to the pseudo-equilibrium constant $(=17K)$. Incorporating the effects of activity coefficients of Mn^{++} and HCO_3^- and concentration scales within K, and solving for $C_{Mn^{++}}$:

$$C_{Mn^{++}} = \frac{K_c^* a_{H^+}}{C_{HCO_3^-}}. \tag{8-30}$$

From pH measurement it was found that in any given sediment it could be reasonably assumed that a_{H^+} was constant with depth. The concentration of HCO_3^- was measured in all sediment samples and for each core could be well described (see Figure 8–5) by the empirical expression:

$$C_{HCO_{3\bar{0}}} = C_{HCO_{3\bar{0}}} + L_0[1 - \exp(-L_1 x)], \tag{8-31}$$

where $C_{HCO_{3\bar{0}}}$ refers to the bicarbonate concentration at the sediment-water interface and L_0 and L_1 are constants obtained by curve fitting. By substituting values of $C_{HCO_3^-}$ and a_{H^+} into equation (8-31), the concentration of $C_{Mn^{++}}$ could be calculated at all depths.

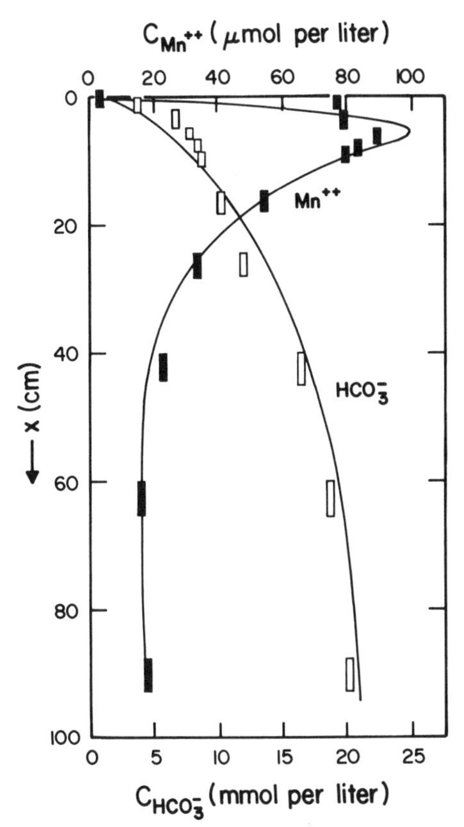

FIGURE 8-5. Concentrations of Mn^{++} and HCO_3^- in sediments of northern Chesapeake Bay (Station 848F) fitted by theoretical curves. (Adapted from Holdren, 1977.)

In order to comply with his overall model, Holdren fitted equation (8-28) to concentration data down to the maximum in Mn^{++} and equation (8-30) to the data below this. As a result of this simplified treatment the depth of maximum concentration represents a discontinuity in the C vs x plot. A theoretically calculated curve is shown fitted to measured pore water Mn^{++} data in Figure 8-5. The best fit of the equilibrium model was obtained for $K_c^* = 15$, which represents an approximately seventeen-fold supersaturation with respect to rhodochrosite.

A major problem with Holdren's model is that the distribution of solid Mn actually found in the same Chesapeake Bay sediments in no way resembles that calculated from the model. This arises mostly from non-steady state fluctuations in the concentration of total Mn delivered to the site of deposition. Such fluctuations need not preclude the use of a steady state model to describe dissolved Mn^{++} (see the discussion of non-steady state effects in Chapter 7); however, the discernment of diagenetic effects in the solids is thereby rendered difficult, due to masking by the fluctuations. Separation of diagenetically active Mn, in the separate forms of rhodochrosite and reducible fine-grained MnO_2, from non-reactive Mn might alleviate this problem but, unfortunately, numerous attempts to do this have been unsuccessful (Holdren, personal communication). At any rate, Holdren's Mn model shows how authigenic mineral formation can be described in terms of equilibrium modeling, an approach that may prove to be useful when applied to other elements and for other sedimentary situations.

Hypersaline Sediments

A striking degree of authigenic mineral formation, alteration, dissolution, and so forth occurs in supersaline sediments (e.g., zeolites, dolomite), but unfortunately there is a general lack of pore water data that would permit proper diagenetic modeling. In addition, because of rapid change in overlying water concentration with time in most smaller salt lakes and playas, which is brought about by fluctuations in rainfall and evaporation, we are often confronted with non-steady state conditions that cannot be adequately described. Studies have been made of short-term salt diffusion in hypersaline sediments (Lerman and Jones, 1973), and our discussion here will be restricted to this topic.

Lerman and Jones (1973) have constructed a diagenetic diffusion model to describe the diffusion of total salts from sediments to overlying water in Lake Abert, Oregon. Apparently the lake water became freshened by accelerated input of stream water 25 years before it was examined by Lerman and Jones. They determined pore water salinities vs depth, which, when combined with diagenetic modeling, were used to describe the increase in salt content of the lake water over the past 25 years by diffusion of dissolved salt out of the underlying sediment.

The model adopted is one of a sudden derease in salinity of the lake (which is uniformly well mixed) at $t = 0$, and subsequent diffusion of salts from the sediment to the overlying water for a period of 25 years. Mathematically, the initial condition and boundary conditions are:

At $t = 0$:

$$x = 0, C = C_L(0),$$

$$x > 0, C = C_\infty;$$

At $t > 0$:

$$x = 0, C = C_L(t),$$

$$x \to \infty, C \to C_\infty,$$

where C refers to concentration of total salts in the sediment pore water and C_L to concentration in the lake water. This is all illustrated in Figure 8-6. In addition to diffusion, the possibility of upward flow of interstitial water into the lake, due to an externally impressed hydrostatic head, was considered. Compaction, bioturbation, burial advection, adsorption, and chemical reaction were assumed to be negligible. Under these conditions the proper diagenetic expression for the sediment is:

$$\frac{\partial C}{\partial t} = D_s \frac{\partial^2 C}{\partial x^2} - v \frac{\partial C}{\partial x}. \tag{8-32}$$

(Here, remember, v refers to water flow only and not to deposition.)

Solution of (8-32) for the given initial and boundary conditions was done analytically, and two rather long expressions, involving exponentials and error functions, were obtained. One expression was for C as a function of depth and time and the other for C_L as a function of time. The interested reader is referred to the original paper of Lerman and Jones for the exact equations. Plots of the expression for pore water concentration vs depth at $t = 25$ years for different values of v and D_s are shown in Figure 8-6. As can be seen, a good fit to the measured data, including the salt concentration in the overlying water, is made by the curve for very low v (0.01 cm yr^{-1}) and $D_s = 1 \times 10^{-6}$ cm^2 sec^{-1}. In other words,

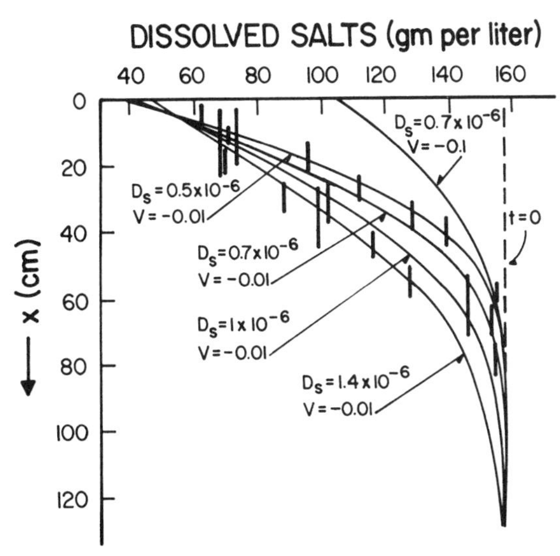

FIGURE 8-6. Distribution of total dissolved salts vs depth in sediments of Lake Abert, Oregon. Also shown are theoretical curves for different values of D_s (in cm^2 sec^{-1}) and v (in cm yr^{-1}). (After Lerman and Jones, 1973.)

the salt concentrations, both in the sediment and in the lake water, are best explained in terms of molecular diffusion without flow of pore water. This is reasonable since the sediments are fine-grained clays and, therefore, should be rather impermeable. The somewhat low D_s value needed to fit the measured pore water concentrations may be explainable in terms of a small degree of dissolution of salts in the sediment. (Major species are Na^+, Cl^-, and HCO_3^-, and their diffusion coefficients in a surficial sediment would be expected to be about three times higher.) By rapidly resupplying salt lost by diffusion, equilibrium dissolution would manifest itself, just like desorption, in terms of a lowered diffusion coefficient.

References

Adamson, A. W., 1967, *Physical Chemistry of Surfaces*, 2nd ed., Wiley-Interscience, New York, 747 p.

Allen, J.R.L., 1970, *Physical Processes of Sedimentation*, American Elsevier, N.Y., 248 p.

Aller, R. C., 1977, The influence of macrobenthos on chemical diagenesis of marine sediments, unpublished Ph.D. dissertation, Yale Univ., 600 p.

Aller, R. C., 1980, Diagenetic processes near the sediment-water interface of Long Island Sound, I. Decomposition and nutrient element geochemistry (S, N, P), in *Estaurine Physics and Chemistry: Studies in Long Island Sound* (ed. B. Saltzman), Advances in Geophysics, v. 22, Academic Press (in press).

Andersen, N. R. and Malahoff, A., eds., 1977, *The Fate of Fossil Fuel* CO_2 *in the Oceans*, Plenum, New York, 749 p.

Anderson, D. E. and Graf, D. L., 1978, Ionic diffusion in naturally-occurring aqueous solutions: use of activity coefficients in transition-state models, *Geochim.* et *Cosmochim. Acta*, v. 42, pp. 251–262.

Aris, R., 1975, *The Mathematical Theory of Diffusion and Reaction in Permeable Catalysts*, Clarendon Press, Oxford, 444 p.

Barnes, R. O. and Goldberg, E. D., 1976, Methane production and consumption in anoxic marine sediments, *Geology*, v. 4, pp. 297–300.

Bathurst, R.G.C., 1971, *Carbonate Sediments and Their Diagenesis*, Elsevier, Amsterdam, 620 p.

Bear, J., 1972, *Dynamics of Fluids in Porous Media*, American Elsevier, New York, 764 p.

Benninger, L. K., Aller, R. C., Cochran, J. K., and Turekian, K. K., 1979, Effects of biological sediment mixing on the ^{210} Pb chronology and trace metal distribution in a Long Island Sound sediment core, *Earth Planet. Sci. Letters*, v. 43, pp. 241–259.

Ben-Yaakov, S., 1972, Diffusion of sea water ions I. Diffusion of sea water into a dilute solution, *Geochim. et Cosmochim. Acta*, v. 36, pp. 1395–1406.

Berger, W. H., 1970, Planktonic foraminifera: selective solution and the lysocline, *Mar. Geology*, v. 8, pp. 111–138.

Berger, W. H., 1976, Biogeneous deep sea sediments: production, preservation and interpretation, in *Chemical Oceanography*, v. 5 (eds. J. P. Riley and R. Chester), Academic Press, New York, pp. 265–387.

Berger, W. H. and Heath, G. R., 1968, Vertical mixing in pelagic sediments, *Jour. Mar. Research*, v. 26, pp. 134–143.

Bergmann, W., 1963, Geochemistry of lipids, in *Organic Geochemistry* (ed. I. A. Breger), Macmillan, New York, pp. 503–542.

Berner, R. A., 1964, An idealized model of dissolved sulfate distribution in recent sediments, *Geochim. et Cosmochim. Acta*, v. 28, pp. 1497–1503.

Berner, R. A., 1968, Rate of concretion growth, *Geochim. et Cosmochim. Acta*, v. 32, pp. 477–483.

Berner, R. A., 1969a, Chemical changes affecting dissolved calcium during the bacterial decomposition of fish and clams in sea water, *Mar. Geology*, v. 7, pp. 253–274.

Berner, R. A., 1969b, Migration of iron and sulfur within anaerobic sediments during early diagenesis, *Am. Jour. Sci.*, v. 267, pp. 19–42.

Berner, R. A., 1971, *Principles of Chemical Sedimentology*, McGraw-Hill, New York, 240 p.

Berner, R. A., 1974, Kinetic models for the early diagenesis of nitrogen, sulfur, phosphorous, and silicon in anoxic marine sediments, in *The Sea*, v. 5 (ed. E. D. Goldberg), Wiley, New York, pp. 427–450.

Berner, R. A., 1975, Diagenetic models of dissolved species in the interstitial waters of compacting sediments, *Am. Jour. Sci.*, v. 275, pp. 88–96.

Berner, R. A., 1976, Inclusion of adsorption in the modelling of early diagenesis, *Earth Planet. Sci. Letters*, v. 29, pp. 333–340.

Berner, R. A., 1977, Stoichiometric models for nutrient regeneration in anoxic sediments, *Limnol. and Oceanog.*, v. 22, pp. 781–786.

Berner, R. A., 1978a, Sulfate reduction and the rate of deposition of marine sediments, *Earth Planet. Sci. Letters*, v. 37, pp. 492–498.

Berner, R. A., 1978b, Rate control of mineral dissolution under earth surface conditions, *Am. Jour. Sci.*, v. 278, pp. 1235–1252.

Berner, R. A., 1979, Kinetics of nutrient regeneration in anoxic marine sediments, in *Origin and Distribution of the Elements*, v. 2 (ed. L. H. Ahrens), Pergamon, Oxford, pp. 279–292.

Berner, 1980, A rate model for organic matter decomposition during bacterial sulfate reduction in marine sediments, in *Biogeochemistry of Organic Matter at the Sediment-Water Interface*, Comm. Natl. Recherche Scientifique (France) Reports (in press).

Berner, R. A. and Morse, J. W., 1974, Dissolution kinetics of calcium carbonate in seawater: IV. Theory of calcite dissolution, *Am. Jour. Sci.*, v. 274, pp. 108–134.

Berner, R. A., Scott, M. R., and Thomlinson, C., 1970, Carbonate alkalinity in the pore waters of anoxic marine sediments, *Limnol. and Oceanog.*, v. 15, pp. 544–549.

Berner, R. A., Westrich, J. T., Graber, R., Smith, J., and Martens, C. S., 1978, Inhibition of aragonite precipitation from supersaturated seawater: a laboratory and field study, *Am. Jour. Sci.*, v. 278, pp. 816–837.

Blatt, H., Middleton, G., and Murray, R., 1972, *Origin of Sedimentary Rocks*, Prentice-Hall, Englewood Cliffs, N.J., 342 p.

Boer, R. B. de, 1977, On the thermodynamics of pressure solution-interaction between chemical and mechanical forces, *Geochim. et Cosmochim. Acta*, v. 41, pp. 249–256.

Boudreau, B. P. and Guinasso, N. L., 1980, The influence of a diffusive sublayer on accretion, dissolution and diagenesis at the sea floor, in *The Dynamic Environment of the Ocean Floor* (eds. K. A. Fanning and F. T. Manheim) D. C. Heath (in press).

Bricker, O. P., 1971, *Carbonate Cements*, Johns Hopkins Press, Baltimore, 376 p.

Broecker, W. S., 1971, A kinetic model for the chemical composition of seawater, *Quaternary Research*, v. 1, pp. 188–207.

Burst, J. F., 1976, Argillaceous sediment dewatering, *Ann. Rev. Earth Planet. Sci.*, v. 4, pp. 293–318.

Burton, W. K., Cabrera, N., and Frank, F. C., 1951, The growth of crystals and the equilibrium structure of their surfaces, *Royal Soc. London Philos. Trans.*, v. A-243, pp. 299–358.

Chave, K. E. and Suess, E., 1970, Calcium carbonate saturation in seawater: effects of dissolved organic matter, *Limnol. and Oceanog.*, v. 15, pp. 633–637.

Christoffersen, J., Christoffersen, M. R., and Kjaergaard, N., 1978, The kinetics of dissolution of calcium hydroxyapatite in water at constant pH, *Jour. Crystal Growth*, v. 43, pp. 501–511.

Claypool, G. and Kaplan, I. R., 1974, The origin and distribution of methane in marine sediments, in *Natural Gases in Marine Sediments* (ed. I. R. Kaplan), Plenum, New York, pp. 99–139.

Cochran, J. K., 1980, The geochemistry of ^{226}Ra and ^{228}Ra in marine sediments, *Am. Jour. Sci.*, v. 280 (in press).

Cornish-Bowden, A., 1976, *Principles of Enzyme Kinetics*, Butterworths, London, 206 p.

Crank, J., 1975, *The Mathematics of Diffusion*, 2nd ed., Pergamon Press, Oxford, 414 p.

Degens, E. T., 1965, *Geochemistry of Sediments*, Prentice-Hall, Englewood Cliffs, N.J., 634 p.

de Kanel, J. and Morse, J. W., 1978, The chemistry of orthophosphate uptake from seawater onto calcite and aragonite, *Geochim. et Cosmochim. Acta.*, v. 42, pp. 1335–1340.

Devol, A. H., 1978, Bacterial oxygen uptake kinetics as related to biological processes in oxygen-deficient zones of the oceans, *Deep Sea Research*, v. 25, pp. 137–146.

Dixon, M. and Webb, E. C., 1964, *Enzymes*, Academic Press, New York, 950 p.

Dobrovolsky, E. V. and Lyalko, V. I., 1979, Dynamic model of fluorite mineralization of carbonate-bearing rocks, *Am. Jour. Sci.*, v. 279, pp. 1022–1032.

Domenico, P. A., 1972, *Concepts and Models in Groundwater Hydrology*, McGraw-Hill, New York, 406 p.

Drever, J. I., 1974, The magnesium problem, in *The Sea*, v. 5 (ed. E. D. Goldberg), Wiley, New York, pp. 337–358.

Duursma, E. K. and Bosch, C. J., 1970, Theoretical, experimental, and field studies concerning diffusion of radioisotopes in sediments and suspended particles of the sea, Part B: Methods and experiments, *Netherlands Jour. of Sea Research*, v. 4, pp. 395–469.

Duursma, E. K. and Eisma, D., 1973, Theoretical, experimental, and field studies concerning reactions of radioisotopes with sediments and suspended particles of the sea, Part C: Applications to field studies, *Netherlands Jour. of Sea Research*, v. 6, pp. 265–324.

Duursma, E. K. and Hoede, C., 1967, Theoretical, experimental, and field studies concerning molecular diffusion of radioisotopes in sediments and suspended solid particles of the sea, Part A: Theories and mathematical calculation, *Netherlands Jour. of Sea Research*, v. 3, pp. 423–457.

Edmond, J. M. and Gieskes, J. M., 1970, On the calculation of the degree of saturation of sea water with respect to calcium carbonate under in situ conditions, *Geochim. et Cosmochim. Acta*, v. 34, pp. 1261–1292.

Emerson, S., and Widmer, G., 1978, Early diagenesis in anaerobic lake sediments II. Thermodynamic and kinetic factors controlling the formation of iron phosphate, *Geochim. et Cosmochim. Acta*, v. 42, pp. 1307–1316.

Emery, K. O., 1960, *The Sea Off Southern California*, Wiley, New York, 366 p.

Engelhardt, W. V., 1977, *The Origin of Sediments and Sedimentary Rocks*, Halstead Press, New York, 359 p.

Engelhardt, W. V. and Gaida, K. H., 1963, Concentration changes of pore solutions during the compaction of clay sediments, *Jour. Sed. Petrology*, v. 33, pp. 919–930.

Farrow, H. T., 1953, The melting of ice cubes in alcohol, *Jour Appl. Biogeochemistry*, v. 17, pp. 314–322.

Filipek, L. H. and Owen, R. M., 1980. Early diagenesis of organic carbon and sulfur in outer shelf sediments from the Gulf of Mexico, *Am. Jour. Sci.*, v. 280 (in press).

Fisher, G. W., 1978, Rate laws in metamorphism, *Geochim. et Cosmochim. Acta*, v. 42, pp. 1035–1050.

Flicker, M. and Ross, J., 1974, Mechanism of chemical instability for periodic precipitation phenomena, *Jour. Chem. Phys.*, v. 60, pp. 3458–3465.

Folk, R. L., 1974, *Petrology of Sedimentary Rocks*, Hemphill, Austin, Texas, 182 p.

Frank, F. C., 1950, Radial symmetric phase growth controlled by diffusion, *Royal Soc. London Proc.*, v. A-201, pp. 48–54.

Fritz, B., 1975, Étude thermodynamique et simulation des réactions entre mineraux et solutions. Application a la géochimie des altérations et des eaux continentales, Université Louis Pasteur de Strasbourg, Institut de Geologie, Mem. 41, 153 p.

Froelich, P. N., Klinkhammer, G. P., Bender, M. L., Luedtke, N. A., Heath, G. R., Cullen, D., Dauphin, P., Hammond, D., Hartman, B., and Maynard, V., 1979, Early oxidation of organic matter in pelagic sediments of the eastern equatorial Atlantic: suboxic diagenesis, *Geochim. et Cosmochim. Acta*, v. 43, pp. 1075–1090.

Fouillac, C., Michard, G., and Bocquier, G., 1977, Une méthode de simulation de l'évolution des profils d'altération, *Geochim. et Cosmochim. Acta*, v. 41, pp. 207–213.

Füchtbauer, H., 1974, *Sediments and Sedimentary Rocks, Part II*, Halstead Press, New York, 464 p.

Garrels, R. M. and Christ, C. L., 1965, *Solutions, Minerals and Equilibrium*, Harper, New York, 450 p.

Glasby, G. P., 1973, Interstitial waters in marine and lacustrine sediments: review, *Jour. Roy. Soc. New Zealand*, v. 3, pp. 43–59.

Goldberg, E. D. and Koide, M., 1962, Geochronological studies of deep-sea sediments by the Io/Th method, *Geochim. et Cosmochim. Acta*, v. 26, pp. 417–450.

Goldhaber, M. B., Aller, R. C., Cochran, J. K., Rosenfeld, J. K., Martens, C. S., and Berner, R. A., 1977, Sulfate reduction, diffusion, and bioturbation in Long Island Sound sediments: report of the FOAM group, *Am. Jour. Sci.*, v. 277, pp. 193–237.

Goldhaber, M. B. and Kaplan, I. R., 1974, The sulfur cycle, in *The Sea*, v. 5 (ed. E. D. Goldberg), Wiley, New York, pp. 569–655.

Goreau, T. J., 1977, Quantitative effects of sediment mixing on stratigraphy and biogeochemistry: a signal theory approach, *Nature*, v. 265, pp. 525–526.

Guinasso, N. L. and Schink, D. R., 1975, Quantitative estimate of biological mixing rates in abyssal sediments, *Jour. Geophys. Research*, v. 80, pp. 3032–3043.

Hakanson, L. and Kallstrom, A., 1978, An equation of state for biologically active lake sediments and its implications for interpretations of sediment data, *Sedimentology*, v. 25, pp. 205–226.

Hamilton, E. L., 1976, Variations of density and porosity with depth in deep-sea sediments, *Jour. Sed. Petrology*, v. 46, pp. 280–300.

Hanor, J. S. and Marshall, J. S., 1971, Mixing of sediment by organisms, in *Trace Fossils*, (ed. Bob F. Perkins), Louisiana State University Miscellaneous Publication 71-1, pp. 127–135.

Hartmann, M., Müller, P., Suess, E., and ven der Weijden, C. H., 1973, Oxidation of organic matter in recent marine sediments, *Meteor. Forschungs. Ergeb.*, Reihe c, v. 12, pp. 74–86.

Heath, G. R., Moore, T. C., and Dauphin, J. P., 1977, Organic carbon in deep sea sediments, in *The Fate of Fossil Fuel* CO_2 *in the Oceans*, (eds. N. R. Andersen and A. Malahoff), Plenum, New York, pp. 605–625.

Heathershaw, A. D., 1974, "Bursting" phenomena in the sea, *Nature*, v. 248, pp. 394–395.

Helfferich, F., ed., 1962, *Ion Exchange*, McGraw-Hill, New York, 624 p.

Helgeson, H. C., Delany, J. M., Nesbitt, H. W., and Bird, D. K., 1978, Summary and critique of the thermodynamic properties of rock-forming minerals, *Am. Jour. Sci.*, v. 278-A, 229 p.

Helgeson, H. C., Garrels, R. M., and Mackenzie, F. T., 1969, Evaluation of irreversible reactions in geochemical processes involving minerals and aqueous solutions—II. Application, *Geochim. et Cosmochim. Acta*, v. 33, pp. 455–481.

Holdren, G. R., 1977, Distribution and behavior of manganese in the interstitial waters of Chesapeake Bay sediments during early diagenesis, unpublished Ph.D. dissertation, Johns Hopkins Univ., Baltimore, Md., 190 p.

Holdren, G. R., Bricker, O. P., and Matisoff, G., 1975, A model for the control of dissolved manganese in the interstitial waters of Chesapeake Bay, in *Marine Chemistry in the Coastal Environment* (ed. T. M. Church), Washington, Am. Chem. Soc. series no. 18, pp. 364–381.

Hurd, D. C., 1972, Factors affecting dissolution rate of biogenic opal in seawater, *Earth Planet. Sci. Letters*, v. 15, pp. 411–417.

Hurd, D. C., 1973, Interactions of biogenic opal, sediment, and seawater in the Central Equatorial Pacific, *Geochim. et Cosmochim. Acta*, v. 37, pp. 2257–2282.

Imboden, D. M., 1975, Interstitial transport of solutes in non-steady state accumulating and compacting sediments, *Earth Planet. Sci. Letters*, v. 27, pp. 221–228.

Jenne, E. A., 1968, Controls on Mn, Fe, Co, Ni, Cu, and Zn concentrations in soils and waters: the significant role of hydrous Mn and Fe oxides, in *Trace Inorganics in Water*, Advances in Chemistry Series 73, pp. 337–387.

Johnson, K. S. and Pytkowicz, R. M., 1979, Ion association of Cl^- with H^+, Na^+, K^+, Ca^{2+}, and Mg^{2+} in aqueous solutions at 25°C, *Am. Jour. Sci.*, v. 278, pp. 1428–1447.

Johnson, T. C., 1975, The dissolution of siliceous microfossils in deep-sea sediments, unpublished Ph.D. dissertation, Univ. of California, San Diego, 163 p.

Jørgensen, B. B., 1978, comparison of methods for the quantification of bacterial sulfate reduction in coastal marine sediments II. Calculation from mathematical models, *Geomicrobiology Jour.*, v. 1, pp. 20–47.

Kaplan, I. R., Emery, K. O., and Rittenberg, S. C., 1963, The distribution and isotopic abundance of sulphur in recent marine sediments off southern California, *Geochim. et Cosmochim. Acta*, v. 27, pp. 297–331.

Keir, R. A., 1979, The dissolution kinetics of biogenic calcium carbonate: Laboratory measurements and geochemical implications, unpublished Ph.D. dissertation, Yale Uni., 284 p.

Kononova, M. M., 1966, *Soil Organic Matter*, 2nd English ed., Pergamon, New York, 554 p.

Krom, M. D. and Berner, R. A., 1980, The experimental determination of the diffusion coefficients of sulfate, ammonium, and phosphate in anoxic marine sediments, *Limnol. and Oceanog.*, v. 25, pp. 327–337.

Laidler, K. J., 1965, *Chemical Kinetics*, McGraw-Hill, New York, 566 p.

Langmuir, D., 1971, Particle size effect on the reaction goethite = hematite + water, *Am. Jour. Sci.*, v. 271, pp. 147–156.

Lasaga, A. C., 1979, The treatment of multi-component diffusion and ion pairs in diagenetic fluxes, *Am. Jour. Sci.*, v. 279, pp. 324–346.

Lasaga, A. C. and Holland, H. D., 1976, Mathematical aspects of non-steady state diagenesis, *Geochim. et Cosmochim. Acta*, v. 40, pp. 257–266.

Lasserre, P., 1976, Metabolic activities of benthic microfauna and meiofauna: recent advances and review of suitable methods of analysis, in *The Benthic Boundary Layer* (ed. I. N. McCave), Plenum Press, New York, pp. 95–142.

Latimer, W. M., 1952, *Oxidation Potentials*, Prentice-Hall, Englewood Cliffs, N. J., 392 p.

Lawson, D. S., Hurd, D. C., and Pankratz, H. S., 1978, Silica dissolution rates of decomposing phytoplankton assemblages at various temperatures, *Am. Jour. Sci.*, v. 278, pp. 1373–1393.

Lerman, A., 1975, Maintenance of steady state in oceanic sediments, *Am. Jour. Sci.*, v. 275, pp. 609–635.

Lerman, A., 1979, *Geochemical Processes: Water and Sediment Environments*, Wiley, New York, 481 p.

Lerman, A. and Jones, B. F., 1973, Transient and steady-state transport between sediments and brine in closed lakes, *Limnol. and Oceanog.*, v. 18, pp. 72–85.

Lerman, A. and Weiler, R. R., 1970, Diffusion of chloride and sodium in Lake Ontario sediments, *Earth Planet. Sci. Letters*, v. 10, pp. 150–156.

Levich, V. G., 1962, *Physicochemical Hydrodynamics*, Prentice-Hall, N.J., 700 p.

Li, Y-H., and Gregory, S., 1974, Diffusion of ions in sea water and in deep-sea sediments, *Geochim. et Cosmochim. Acta*, v. 38, pp. 703–714.

Lippman, F., 1973, *Sedimentary Carbonate Minerals*, Springer-Verlag, New York-Heidelberg, 288 p.

Manheim, F. T., 1970, The diffusion of ions in unconsolidated sediments, *Earth Planet. Sci. Letters*, v. 9, pp. 307–309.

Manheim, F. T., 1976, Interstitial waters of marine sediments, in *Chemical Oceanagraphy*, v. 6 (eds. J. P. Riley and G. Skirrow), pp. 114–186.

Manheim, F. T. and Waterman, L. S., 1974, Diffusimitry (diffusion constant estimation) on sediment cores by resistivity probe, *Initial Reports of the Deep Sea Drilling Project*, v. 22, pp. 663–670.

Martens, C. S. and Berner, R. A., 1977, Interstitial water chemistry of anoxic Long Island Sound sediments 1. Dissolved gases, *Limnol. and Oceanog.*, v. 22, pp. 10–25.

Martens, C. S., Berner, R. A., and Rosenfield, J. K., 1978, Interstitial water chemistry of anoxic Long Island Sound sediments, 2. Nutrient regeneration and phosphate removal, *Limnol. and Oceanog.*, v. 23, pp. 605–617.

Martens, C. S. and Harriss, R. C., 1970, Inhibition of apatite precipitation in the marine environment by mangnesium ions, *Geochim. et Cosmochim. Acta*, v. 34, pp. 621–625.

Matisoff, G., 1980, Early diagenesis of Chesapeake Bay sediments: Part 1.

A time series study of temperature and chloride, *Am. Jour. Sci.*, v. 280, pp. 1–25.

McDuff, R. E. and Ellis, R. A., 1979, Determining diffusion coefficients in marine sediments: a laboratory study of the validity of resistivity techniques, *Am. Jour. Sci.*, v. 279, pp. 666–675.

McDuff, R. E. and Gieskes, J. M., 1976, Calcium and magnesium profiles in DSDP intersititai waters: diffusion or reaction?, *Earth Planet. Sci. Letters*, v. 33, pp. 1–10.

Meade, R. H., 1966, Factors influencing the early stages of the compaction of clays and sands, *Jour. Sed. Petrology*, v. 36, pp. 1085–1101.

Menzel, D. W., 1974, Primary productivity, dissolved and particulate organic matter, and the sites of oxidation of organic matter, in *The Sea*, v. 5 (ed. E. D. Goldberg) Wiley, New York, pp. 659–678.

Michard, G., 1971, Theoretical model for manganese distribution in calcareous sediment cores, *Jour. Geophys. Research*, v. 76, pp. 2179–2186.

Millero, F. J., 1979, *Effects of Pressure and Temperature on Activity Coefficients in Electrolyte Solutions*, (ed. R. M. Pytkowicz), CRC Press, Palm Beach, Fla., (in press).

Morse, J. W., 1974, Calculation of diffusive fluxes across the sediment-water interface, *Jour. Geophys. Research*, v. 33, pp. 5045–5048.

Morse, J. W., 1978, Dissolution kinetics of calcium carbonate in sea water: VI. The near equilibrium dissolution kinetics of calcium carbonate-rich deep sea sediments, *Am. Jour. Sci.*, v. 278, pp. 344–353.

Morse, J. W. and Berner, R. A., 1979, The chemistry of calcium carbonate in the deep oceans, in *Chemical Modeling—Speciation, Sorption, Solubility and Kinetics in Aqueous Systems*, (ed. E. A. Jenne), Am. Chem. Soc. Symposium Series, No. 93, pp. 499–535.

Murray, J. W., Grundmanis, V., and Smethie, W. M., 1978, Interstitial water chemistry in the sediments of Saanich Inlet, *Geochim. et Cosmochim. Acta*, v. 42, pp. 1011–1026.

Murthy, A.S.P. and Ferrell, R. E., 1972, Comparative chemical composition of sediment interstitial waters, *Clays and Clay Minerals*, v. 20, pp. 317–321.

Nicolas, G. and Portnow, J., 1973, Chemical oscillations, *Chem. Rev.*, v. 73, pp. 365–384.

Nielsen, A. E., 1961, Diffusion controlled growth of a moving sphere. The kinetics of crystal growth in potassium perchlorate precipitation, *Jour. Phys. Chem.*, v. 65, pp. 46–49.

Nielsen, A. E., 1964, *Kinetics of Precipitation*, MacMillan, New York, 151 p.

Nielsen, A. E. and Söhnel, O., 1971, Interfacial tensions electrolyte crystal-aqueous solution from nucleation data, *Jour. Cryst. Growth*, v. 11, pp. 233–242.

Nozaki, Y., Cochran, J. K., Turekian, K. K., and Keller, G., 1977, Radiocarbon and ^{210}Pb distribution in submersible-taken deep-sea cores from Project Famous, *Earth Planet. Sci. Letters*, v. 34, pp. 167–173.

Nriagu, J. O. and Dell, C. I., 1974, Diagenetic formation of iron phosphates in recent lake sediments, *Am. Mineralogist*, v. 59, pp. 934–946.

Ohara, M. and Reid, R. C., 1973, *Modeling Crystal Growth Rates from Solution*, Prentice-Hall, Englewood Cliffs, N.J., 272 p.

Ostroumov, E. A., Volkov, I. I., and Fomina, L. C., 1961, Distribution of sulfur compounds in bottom sediments of the Black Sea, *Trudy Inst. Okeanol.*, Akad. Nauk. S.S.S.R., v. 50, pp. 93–129 (in Russian).

Peng, T.-H., Broecker, W. S., Kipphut, G., and Shackleton, N., 1977, Benthic mixing in deep sea cores as determined by ^{14}C dating and its implications regarding climate, stratigraphy, and the fate of fossil fuel CO_2, in *The Fate of Fossil Fuel* CO_2 *in the Oceans*, (eds. N. R. Andersen and A. Malahoff), Plenum, New York, pp. 355–374.

Perry, E. A., Gieskes, J. M., and Lawrence, J. R., 1976, Mg, Ca and O^{18}/O^{16} exchange in the sediment-pore water system, Hole 149, DSDP, *Geochim. et Cosmochim. Acta*, v. 40, pp. 413–423.

Peterson, M.N.A. and Griffin, J. J., 1964, Volcanism and clay minerals in the southeastern Pacific, *Jour. Mar. Research*, v. 22, pp. 13–21.

Pettijohn, F. J., 1975, *Sedimentary Rocks*, 3rd ed., Harper, New York, 628 p.

Pettijohn, F. J., Potter, P. E., and Siever, R., 1972, *Sand and Sandstone*, Springer-Verlag, New York-Heidelberg, 618 p.

Pytkowicz, R. M., 1975, Activity coefficients of carbonates and bicarbonates in seawater, *Limnol. and Oceanog.*, v. 30, pp. 971–975.

Redfield, A. C., 1958, The biological control of chemical factors in the environment, *Am. Sci.*, v. 46, pp. 205–222.

Reeburgh, W. S. and Heggie, D. T., 1974, Depth distributions of gases in shallow water sediments, in *Natural Gases in Marine Sediments* (ed. I. R. Kaplan), Plenum, New York, pp. 27–45.

Rhoads, D. C., 1974, Organism-sediment relations on the muddy sea floor, *Oceanogr. Mar. Biol. Am. Rev.*, v. 12, pp. 263–300.

Rhoads, D. C. and Morse, J. W., 1971, Evolutionary and ecologic significance of oxygen-deficient marine basins, *Lethaia*, v. 4, pp. 413–428.

Rickard, D. T., 1975, Kinetics and mechanism of pyrite formation at low temperatures, *Am. Jour. Sci.*, v. 274, pp. 636–652.

Rieke, H. H. and Chilingarian, G. V., 1974, *Compaction of Argillaceous Sediments*, Elsevier, Amsterdam, 424 p.

Robbins, J. A. and Callender, E., 1975, Diagenesis of manganese in Lake Michigan sediments, *Am. Jour. Sci.*, v. 275, pp. 512–533.

Robie, R. S., Hemingway, B. S., and Fisher, J. R., 1978, Thermodynamic properties of minerals and related substances at 298.15°K and 1 bar (10^5 pascals) pressure and at higher temperatures, *U.S. Geological Survey Bull.* 1452, 456 p.

Robin, P.-Y., 1978, Pressure solution at grain-to-grain contacts, *Geochim. et Cosmochim. Acta*, v. 42, pp. 1383–1389.

Robinson, R. A. and Stokes, R. M., 1959, *Electrolyte Solutions*, Butterworths, London, 571 p.

Rosenfeld, J. K., 1979, Ammonium adsorption in nearshore anoxic sediments, *Limnol. and Oceanog.*, v. 24, pp. 356–364.

Rosenfeld, J.K., 1980, Nitrogen diagenesis in Long Island Sound sediments, *Am. Jour. Sci.*, v. 380 (in press).

Ruddiman, W. F. and Glover, L. K., 1972, Vertical mixing of ice-rafted volcanic ash in North Atlantic sediments, *Bull. Geol. Soc. America*, v. 83, pp. 2817–2836.

Sayles, F. L., 1979, The composition and diagenesis of interstitial solutions I. Fluxes across the seawater-sediment interface in the Atlantic Ocean, *Geochim. et Cosmochim. Acta*, v. 43, pp. 526–546.

Sayles, F. L. and Mangelsdorf, P. C., 1977, The equilibration of clay minerals with seawater: exchange reactions, *Geochim. et Cosmochim. Acta*, v. 41, pp. 951–960.

Schink, D. R., Guinasso, N. L., and Fanning, K. A., 1975, Processes affecting the concentration of silica at the sediment-water interface of the Atlantic Ocean, *Jour. Geophys. Research*, v. 80, pp. 3013–3031.

Schink, D. R., and Guinasso, N. L., 1977, Modelling the influence of bioturbation and other processes on calcium carbonate dissolution at the sea floor, in *The Fate of Fossil Fuel* CO_2 *in the Ocean* (ed. N. R. Andersen and A. Malahoff) Plenum Press, New York, pp. 375–400.

Schink, D. R. and Guinasso, N. L., 1978a, Redistribution of dissolved and adsorbed materials in abyssal marine sediments undergoing biological stirring, *Am. Jour. Sci.*, v. 278, pp. 687–702.

Schink, D. R. and Guinasso, N. L., 1978b, Possible role of aragonite in separating the calcite lysocline from the depth of calcite saturation, *EOS*, v. 59, p. 411.

Schnitzer, M. and Khan, S. U., 1972, *Humic Substances in the Environment*, Dekker, New York, 327 p.

Shishkina, O. V., 1972, *Geocehemistry of Marine and Oceanic Interstitial Waters*, Izdatel'stvo Nauka, Moscow, 227 p. (in Russian).

Simkiss, K., 1964, The inhibitory effects of some metabolites on the precipitation of calcium carbonate from artificial and natural seawater, *Jour. Cons., Cons., Perm. Int. Explor. Mer.*, v. 29, pp. 6–18.

Smith, J. E., 1971, The dynamics of shale compaction and evolution of pore-fluid pressures, *Mathematical Geology*, v. 3, pp. 239–262.

Sorokin, Y. I., 1957, Ability of sulfate reducing bacteria to utilize methane for reduction of sulfates to hydrogen sulfide, *Dokl. Akad. Nauk.*, SSSR, v. 115, pp. 816–818.

Stern, K. H., 1954, The liesegang phenomenon, *Chem. Rev.*, v. 54, pp. 79–99.

Stumm, W, and Leckie, J. O., 1970, Phosphate exchange with sediments: its role in the productivity of surface waters, in *Advances in Water Pollution Research*, v. 2, pp. 26/1–26/16.

Stumm, W. and Morgan, J. J., 1970, *Aquatic Chemistry*, Wiley, New York, 583 p.

Suess, E., 1976, Nutrients near the depositional interface, in *The Benthic Boundary Layer*, (ed. I. N. McCave), Plenum, New York, pp. 57–79.

Suess, E., 1979, Mineral phases formed in anoxic sediments by microbial decomposition of organic matter, *Geochim. et Cosmochim. Acta*, v. 43, pp. 339–352.

Takahashi, T., 1975, Carbonate chemistry of seawater and the calcite compensation depth in the oceans, *Cushman Found. Foram. Research Spec. Publ.* 13, (eds. W. V. Sliter, A.W.H. Be, and W. H. Berger), pp. 11–26.

Takahashi, T. and Broecker, W. S., 1977, Mechanism for calcite dissolution on the sea floor, in *The Fate of Fossil Fuel* CO_2 *in the Oceans*, (eds. N. R. Andersen and A. Malahoff) Plenum, New York, pp. 455–478.

Thimann, K. V., 1963, *The Life of Bacteria* 2nd ed., MacMillan, New York, 909 p.

Toth, D. J. and Lerman, A., 1977, Organic matter reactivity and sedimentation rates in the ocean, *Am. Jour. Sci.*, v. 277, pp. 265–285.

Truesdell, A. H. and Christ, C. L., 1968, Cation exchange in clays interpreted by regular solution theory, *Am. Jour. Sci.*, v. 266, pp. 402–412.

Truesdell, A. H., and Jones, B. F., 1974, WATEQ, a computer program for calculating chemical equilibria of natural waters, *U.S. Geol. Surv. Jour. Research*, v. 2, pp. 233–248.

Turekian, K. K., Cochran, J. K., and DeMaster, D. J., 1978, Bioturbation in deep-sea deposits: rates and consequences, *Oceanus*, v. 21, pp. 34–41.

Turk, J. T., 1976, A study of diffusion in clay-water systems by chemical and electrical methods, unpublished Ph.D. thesis, Univ. of California, San Diego, 111 p.

Twenhofel, W. H., 1961, *Treatise on Sedimentation*, v. 2, Dover, New York, 926 p.

Vanderborght, J.P. and Billen, G., 1975, Vertical distribution of nitrate concentration in interstitial water of marine sediments with nitrification and denitrification, *Limnol. and Oceanog.*, v. 20, pp. 953–961.

Vanderborght, J.P., Wollast, R., and Billen, G., 1977a, Kinetic models of diagenesis in disturbed sediments. Part 1. Mass transfer properties and silica diagenesis, *Limnol. and Oceanog.*, v. 22, pp. 787–793.

Vanderborght, J.P., Wollast, R., and Billen, G., 1977b, Kinetic models of diagenesis in disturbed sediments. Part 2. Nitrogen diagenesis, *Limnol. and Oceanog.*, v. 22, pp. 794–803.

Van Olphen, H., 1977, *An Introduction to Clay Colloid Chemistry*, 2nd ed., Wiley, New York, 318 p.

Vinograd, J. R. and McBain, J. W., 1941, Diffusion of electrolytes and of the ions in their mixtures, *J. Amer. Chem. Soc.*, v. 63, pp. 2008–2015.

Waage, K. M., 1964, Origin of repeated fossilferous concretion layers in the Fox Hills formation, *Kansas Geological Survey Bull.* 169, pp. 541–563.

Walton, H. F., 1959, Ion exchange equilibria, in *Ion Exchange, Theory and Practice* (ed. F. C. Nachod), Academic Press, New York, pp. 3–28.

Welte, D., 1973, Recent advances in organic geochemistry of humic substances and kerogen. A review, in *Advances in Organic Geochemistry* (eds. B. Tissot and F. Bienner), Editions Technip, Paris, pp. 4–13.

Weyl, P. K., 1959, Pressure solution and the force of crystallization—a phenomenological theory, *Jour. Geophys. Research*, v. 64, pp. 2001–2025.

Whitfield, M., 1975, Sea water as an electrolyte solution, in *Chemical Oceanography*, 2nd ed. (eds. J. P. Riley and G. Skirrow), Academic, London, pp. 43–171.

Wollast, R., 1971, Kinetic aspects of the nucleation and growth of calcite from aqueous solutions, in *Carbonate Cements*, (ed. O. P. Bricker) Johns Hopkins Press, pp. 264–273.

Wollast, R., 1974, The silica problem, in *The Sea*, v. 5 (ed. E. D. Goldberg), Wiley, New York, pp. 359–394.

Wood, J. P., 1975, Thermodynamics of brine-salt equilibria—1. The systems $NaCl–KCl–MgCl_2–CaCl_2–H_2O$ and $NaCl–MgSO_4–H_2O$ at 25°C, *Geochim. et Cosmochim. Acta*, v. 39, p. 1147–1163.

Yingst, J., 1978, Patterns of micro- and meiofaunal abundance in marine sediments measured with the adenosine triphosphate assay, *Marine Biology*, v. 47, pp. 41–54.

Index

Library of Congress Cataloging in Publication Data

Berner, Robert A 1935–
Early diagenesis.

(Princeton series in geochemistry)
Bibliography: p.
Includes index.
1. Diagenesis. I. Title. II. Series.
QE571.B47 552′.5 80-7510
ISBN 0-691-08258-8
ISBN 0-691-08260-X (pbk.)